A NUMERATE LIFE

A
NUMERATE
LIFE

A MATHEMATICIAN EXPLORES
THE VAGARIES OF LIFE,
HIS OWN AND PROBABLY YOURS

JOHN ALLEN
PAULOS

Prometheus Books

59 John Glenn Drive
Amherst, New York 14228

Published 2015 by Prometheus Books

Background image © Media Bakery
Cover image © Shutterstock
Cover design by Grace Conti-Zilsberger

Inquiries should be addressed to

Prometheus Books
59 John Glenn Drive
Amherst, New York 14228
VOICE: 716–691–0133
FAX: 716–691–0137
WWW.PROMETHEUSBOOKS.COM

19 18 17 16 15   5 4 3 2 1

Library of Congress Cataloging-in-Publication Data

Paulos, John Allen.
    A numerate life : a mathematician explores the vagaries of life, his and probably yours / by John Allen Paulos.
    pages cm
    Includes bibliographical references and index.
    ISBN 978-1-63388-118-1 (pbk.) — ISBN 978-1-63388-119-8 (e-book)
    1. Paulos, John Allen. 2. Mathematicians—United States—Biography. I. Title.

QA29.P38P38 2015
510.92—dc23
[B]

    2015023515

For my grandsons, Theo, Charlie, and Max

CONTENTS

Introduction: What's It All About?   11

CHAPTER 1: BULLY TEACHER, CHILDHOOD MATH   19

Some Early Estimates, Speculations   19
Pedagogy, Vanquishing Blowhards and
    Opponents, and Monopoly   23
Of Mothers and Collecting Baseball Cards   27
A Further Note on Math, Humor, and My Education   29

**CHAPTER 2: BIAS, BIOGRAPHY, AND WHY WE'RE
ALL A BIT FAR-OUT AND BIZARRE**   35

Bias and Mindsets, Statistics and Biography   35
Despite Normal Appearances, We're All Strange   41
Misapplications of Mathematics to Everyday Life—
    A Caveat   46

CHAPTER 3: AMBITION VERSUS NIHILISM   49

Infinity, Sets, and Immortality   49
Selves and Absurdity   53
The Story of "I"—Neurons, Hallucinations, and Gödel   58

CHAPTER 4: LIFE'S SHIFTING SHAPES   **63**

Primitive Math, Life Trajectories, and Curve Fitting   63
The Environment as a Pinball Machine,
    the Quincunx of Life   69
Biographies and the Texas Sharpshooter   74

CHAPTER 5: MOVING TOWARD THE
        UNEXPECTED MIDDLE   **79**

A Few Touchstone Memories   79
Lunch, Good Night, and My Parents—
    Milwaukee in the 1950s   83
Logic, Jokes, and Adult Life as an Unexpected Punch Line   85
Memories and Benford's Law   89

CHAPTER 6: PIVOTS—PAST TO PRESENT   **93**

Kovalevsky, Prediction, and My Grandmother's
    Petty Larceny   93
Turning Points, Acadia to Kenya   98
Past Accomplishments versus Present Potential   102

CHAPTER 7: ROMANCE AMONG TRANS-HUMANS
        AND US CIS-HUMANS   **105**

Roboromance and the End of Biography   105
Choosing a Spouse, Meeting My Wife, Sheila   108
Romantic Crushes, Bayesian Statistics, and Life   110
Domestic Math: Toilet Seats, Up or Down;
    Movies, Early or Late   112

CHAPTER 8: CHANCES ARE THAT CHANCES ARE   115

If Only . . . Probability and Coincidences,
    Good, Bad, and Ugly   115
Innumeracy, A Mathematician Reads the Newspaper,
    and Their Aftermath   122

CHAPTER 9: LIVES IN THE ERA OF NUMBERS
AND NETWORKS   129

How Many E-Mails, Where Did We Buy That—
    The Quantified Life   129
A Twitterish Approach to Biography   133
Scale and Predictability   140

CHAPTER 10: MY STOCK LOSS, HYPOCRISY,
AND A CARD TRICK   145

My Stock Loss and a Few Pitfalls of Narrative Logic   145
One Cheer for Hypocrisy   151
Kruskal's Card Trick and Common Denouements   154

CHAPTER 11: BIOGRAPHIES: VERSTEHEN
OR SUPERFICIAL   161

Consciousness, Biographies, and Shmata   161
Leah and Daniel, My Grandsons and I   170
Gompertz's Law of Human Mortality and Life Span   172

**CHAPTER 12: TRIPS, MEMORIES, AND
              BECOMING JADED**   **175**

Topology, Travel, and a Thai Taxi Driver   175
Experiencing versus Remembering Selves
    and Autobiographies   179
Peak Experiences, Record Setting, and the
    Path from Jade to Jaded   182
Joining My Father   185

Notes   189

Index   199

Introduction

WHAT'S IT ALL ABOUT?

My interest in mathematics was galvanized by what in retrospect seems to be horrible pedagogy bordering on child abuse. I mention the incident here and will describe it in less lurid tabloid terms a bit later, because it hints at a motivation for both my career in mathematics and my skeptical attitude toward biography and autobiography. I look back and wonder, Is the story true? Is it balanced? Is it representative? The answer is "yes, more or less," but, although I am its protagonist, I still have reservations about its accuracy. I'm sure I'd have even more about your stories.

Whether because of my natural temperament, my training as a mathematician, or a late midlife reckoning, I look on the whole biographical endeavor, my own included, as a dubious one. Even George Washington's signature line about cutting down the cherry tree, "I cannot tell a lie," is probably flapdoodle. More likely he said, "No comment" or "I don't recall the incident" or maybe "The tree was rotten anyway." I tend to scoff when reading that a new biography has revealed that the great So-And-So always did X because (s)he secretly believed Y. I'm not particularly ornery, but I often react to such statements about the alleged actions or beliefs of well-known people with a silent "That's B.S." A more likely reaction if someone makes the claim directly to me is a polite, but pointed "How do you know that?" or even "How could anyone know that?" or, in the case of autobiographies, "How could anyone remember that?"

Memories are often inaccurate or fabricated, perspectives biased, "laws" and assumptions unfounded, contingencies unpredictable; even the very notion of a self is suspect. (But like the nutritionist who secretly enjoys candy and donuts, I've always enjoyed reading [auto]biographies, ranging from James Boswell's *The Life of Samuel Johnson, LL.D.* to Mary Karr's *Liars' Club*.[1])

Given my skepticism of the biographical enterprise, it might seem I've taken a bold and/or foolhardy step to write a quasi-memoir of my own, but *quasi-* here means "not so much."

True to my doubts, what I've written is a meta-memoir, even an anti-memoir. Employing ideas from mathematics (quite broadly and non-technically construed) as well as analytic philosophy and related realms, but requiring no special background in mathematics, I've tried to convey some of the concerns and questions most of us don't, but arguably should, have when reading biographies and memoirs or even when just thinking about our own lives. The "arguably" is the burden of this book; imparting a certain modicum of mathematical understanding and biographical numeracy is its presumptuous goal. (I say presumptuous because of the nebulousness of the notion of biography and the vast variety of different biographies. In a more concrete direction there is the specificity of the book's focus on conventional biographies, mine and probably yours.)

One of the first questions that comes to mind when considering a life is an abstract "What is its average length?" or perhaps a more visceral "How long have I got?" Quite relevant is evolutionary biologist Stephen Jay Gould's article "The Median Isn't the Message," in which he describes his cancer diagnosis and the associated median life span of eight months that it allowed.[2] But the median, of course, is not the mean or simple average of patients' life spans; it is the life span shorter than which half the patients survive and longer than which the other half do. Moreover, the statistical distribution of life spans is right-skewed, meaning many people live considerably longer than the median as did Gould (twenty years). Knowledge of statistics and distributions allayed his anxieties and, more generally, as I'll try to show, mathematical knowledge can shed much-needed light on many other life situations and life stories.

Let me illustrate with a somewhat disguised statistical point.

Whatever else a biography may be, it is usually considered to be a story, the story of a person's life. And probably people's most common response to a story is a tendency to suspend disbelief when reading, hearing, or viewing one in order not to spoil its enjoyment. "Let's pretend. It'll be fun." This mindset is quite opposed to that prevailing in mathematics and science where people typically suspend belief in order not to jump to conclusions until they have compelling evidence. "Wait. Why should we believe that?" These two different approaches are not unrelated to different tolerances for false-positive and false-negative conclusions, which I'll elaborate on later. Not surprisingly, perhaps, the latter show-me tentativeness is the approach I will adopt here. It's in line with the bumper sticker that counsels: Don't believe everything you think.

How did I come to write a book on such disparate realms as mathematics and biographies? After all, fish don't need bicycles, and flashlights don't use solar power, and biographies don't seem to need or use mathematics, hence this extended justification. One element of my biography (or psychology) that disposed me to write this book is that I've always liked the idea of rubbing together incongruous subjects, which seems to me almost a necessary condition for generating creative ideas. At times this habit of rubbing together has earned me a good number of eye rolls, sometimes even a bit of vituperation. People don't always like it when notions or relations they hold dear have reflections in domains such as mathematics that they consider reductive or somehow inappropriate.

That's too bad considering that mathematics is a most productive way of looking at the world. The philosopher Ludwig Wittgenstein once remarked that he looked forward to the day when philosophy disappeared as a subject but all other subjects were approached philosophically. I have a related but weaker wish for mathematics. I certainly don't wish for it to disappear as a subject, but I do wish that it, too, was more widely seen to be an adverb and that its insights and ideas could inform the approach to all other subjects, including biography. With this as a partial motivation, I have over the years written about the connections between mathematics and humor, journalism, the stock market, analytic philosophy, religion, and a number of other topics (but not fish and bicycles). Nonobvious but significant points of correspondence almost always arise if one searches for them.

Here I hope to show that the points of correspondence between mathematics and biography are, despite superficial appearances, quite profound. Carl Sagan, the astronomer, skeptic, and science writer, wrote that we—our DNA, our teeth, our blood—are starstuff, made out of the very same material as the stars. As naturally occurring entities in the universe, we are, in a sense, also "mathstuff"—changing and developing according to mathematically expressible relations, instantiating mathematical notions of all sorts, and illustrating mathematical principles from diverse fields. "Mathstuff," I maintain, is a defensible neologism since patterns are, at least to mathematicians, nonmaterial stuff. It's thus eminently reasonable to try to obtain an understanding of this mathstuff of which, it can be maintained, we and everything else are made. In particular, how do these mathematical patterns express themselves in our life stories?

A less exalted factor in deciding to write this book is perhaps a bit too much self-reflection (a trait I share with all my fellow parttime solipsists). A while ago I had written a few short autobiographical sketches that I liked. Given my predilections, I wondered if a few of these personal vignettes might serve as jumping-off points, each providing a concrete illustration of general mathematically flavored insights about biographies—observations, perceptions, and experiences that would resonate widely.

For an early (and probably not uncommon) arithmetical example, I recall that even as a very young child I silently determined that the story of Santa Claus's exploits just had to be bogus on quantitative grounds alone—all those chimneys and hot chocolates in one night! Even then I wasn't very good at suspending disbelief. And as an adult I've found it all the more natural to wonder about the probability of certain reported events (not involving Santa) really occurring, whether because of purposeful lying or notoriously unreliable memories, or how to roughly quantify the strangeness of an event or the quirk of someone's personality, or what numbers and logic might intimate about both the quotidian happenings and the long arc of a life. So I've given myself the congenial task here of examining the structure of generic memoirs and biographies from a thoroughly skeptical and perhaps occasionally annoying perspective. I subject even the notion of a romantic crush to a mathematical analysis.

As mentioned, in some of my previous books, including *Innumeracy*, *A Mathematician Reads the Newspaper* and *A Mathematician Plays the Stock Market*, I tried to show how math (again, quite broadly construed) can help us understand and analyze certain areas of real life. I employ some of these same basic mathematical ideas here in an attempt to better comprehend our lives, in particular the enhanced, distorted, even imaginary narratives about ourselves that we all effortlessly create. I am repeatedly struck by recurring questions when I reread some of my brief remembrances or recall events from long ago. How did I get here from there— the psychological as well as the physical path? Not novelist Anne Tyler's "wrong life,"[3] but certainly a different and unexpected one. Who was that kid, the one who, strangely to me now, loved playing with toy soldiers, tanks, and ships, and tacked model airplanes to the ceiling over his bed? How have I changed and how have I just gotten older? To what degree have I misremembered or embroidered events? Where'd my story of me come from?

But what is of much more universal interest are the constraints on any life story—mine, yours, his, hers. In particular, what are the ideas from mathematics that might clarify certain aspects of any biography?

How does one tell a life story or, more accurately, selected parts of it?

To what extent is the choice of incidents related likely to be biased—statistically, psychologically, otherwise?

How should we evaluate past decisions (or future ones)?

What kind of plastic, ephemeral, or nominal entity is the self?

What can one say of the general shape or trajectory of a life story?

And what roles do chaos, coincidence, probability, topology, social media such as Twitter, quantitative constraints, and cognitive delusions play in our lives and in their depiction in biographies?

Some of the specific questions addressed herein are:

How might the notion, borrowed from mathematical logic, of nonstandard models of axiom sets be relevant to the predicting of our futures?

How are our lives, in a profound sense, joke-like?

How does nonlinear dynamics explain the narcissism of small differences sometimes cascading into siblings growing into very different people?

How can simple arithmetic put lifelong habits into perspective?

How can higher-dimensional geometry help us see why we're all literally peculiar, far-out?

How can logarithms and exponentials shed light on why we tend to become jaded and bored as we age?

How can probability and card collecting tell us anything about our so-called bucket lists and the contingency of life's turning points?

How come I begin every question with How?

How can algorithmic complexity and Shannon entropy balance past accomplishments and future potential?

How can we find the curve of best fit that captures the path our lives have taken?

I will herein attempt to explain how these and other mathematical constructs say something of significance about biographies as well as all our unchronicled lives.

One obvious obstacle to the writing of somewhat truthful memoirs and biographies already alluded to is that people, especially the authors, tend to lie or at minimum embroider. And it's not just authors, of course. I remember my grandmother telling my grandfather that it was late and that he should stop playing cards with his friends. His reply was always the same, both humorous and slightly paradoxical, "We'll be through in a little while. We still have a few more lies to tell." I'll discuss lies and the paradoxes to which they sometimes give rise in a bit, but they are not the most daunting obstacle.

As oddities from philosophy and psychology make clear, people's perspectives and evaluations of situations and others' behavior not only differ greatly but change over time. For example, the comedian Louie C. K. rants about a fellow plane passenger who became irate because the Wi-Fi on the plane was so primitive—the plane, you know, that aluminum tube that was flying him through the air at five hundred miles per hour and an altitude of seven miles. Holding historical figures to a contemporary standard of political correctness also illustrates the phenomenon. Recently the life of the great physicist Richard Feynman has undergone reevaluation, and he's been adjudged quite sexist by contemporary mores.

I listen to golden oldies pop songs from the '50s and '60s on Sirius radio, songs Feynman might have listened to occasionally. I'm often shocked by the benighted sentiments implicit in the songs and doubly shocked by the way I can ignore the lyrics and still resonate with the songs.

Another personal example: In *Once Upon a Number* I recalled an event from my childhood that illustrates this change of evaluation. As kids, my brother and I would visit our grandparents and regularly make our way around the leafy neighborhood taking turns throwing darts at the large trees that were planted every twenty-five feet or so between the sidewalks and the streets. We kept meticulous score of how many of them we each hit. Once after he had beaten me, I convinced my brother to have this contest in our underwear. He never realized until we'd returned to our grandparents' house that I was wearing swimming trunks under my underwear. In his anger and my gloating, we both accepted that I somehow had appeared less moronic than he during this escapade.

More generally—a phrase that appears often herein as I move from a personal incident or observation to a more abstract or universal rendering—how can we bring some mathematical muscle to the analysis and evaluation of common biographical anecdotes and stories?

I reiterate (reiterating being an occupational hazard of being a professor) that I take a singular approach: to simultaneously explore biography from a personal point of view and from the perspective of a mathematician interested in the no-man's land where my discipline comes within hailing distance of another domain—somewhere between Plato and Play-Doh, math and myth, Pythagoras and Plutarch. (More to the Play-Doh side is the observation that there is a fine line between numerator and denominator.) The book is terse, perhaps cloyingly self-indulgent in places, although very far from a tell-all and admittedly a somewhat weird amalgam.

Its progression will be episodic and nonlinear (although the autobiographical sketches will be roughly chronological), and connected by an introspective, mathematical sensibility leading to brief discussions of relevant theoretical matters, some distilled from previous works (my personal golden oldies). In a rough, approximate way the numbers and the narratives (if you'll allow the alliteration) will alternate throughout the book and, like life, will be a bit of a mishmash. But, as the Czech writer Milan Kundera asked, "Isn't that exactly the definition of biography? An artificial logic imposed on an 'incoherent succession of images'?"[4]

To set the stage, a few simple facts crammed into a very long sentence to begin and moor this quasi-memoir: I was born in 1945; grew up in Chicago and Milwaukee; graduated from the University of Wisconsin; went into the Peace Corps to avoid the draft; returned to Madison where I met my beauteous wife, Sheila, and received my PhD in mathematics; moved to Philadelphia to teach at Temple University; had two wondrous children (who have recently begotten three grandchildren); wrote some books on mathematics, a couple of which were bestsellers;[5] and gradually, as I became less intelligent, became more a writer than a mathematician. I'll begin with a couple of numerical childhood memories far enough in the past so that the golden gauze of nostalgia has settled over them.

Chapter 1

BULLY TEACHER, CHILDHOOD MATH

SOME EARLY ESTIMATES, SPECULATIONS

Postponing discussion of more engaging biographical issues and more intriguing mathematics, I'll begin simply with my early enchantment with numbers, the atoms of the mathematical world. Long before the Count on *Sesame Street*, I loved to count. I counted anything and everything, including, my father told me, the number of little thin cylinders in a pack of his ubiquitous cigarettes, although I doubt he appreciated the cigarettes being all grubbed up with my sticky toddler fingers. I mentioned in the introduction my numerical issues with Santa Claus. A considerate little fellow, I also remember humoring my parents whenever they mentioned him. I wanted to protect them from my guilty knowledge of his nonexistence, and so I feigned belief. (This is a nice illustration of the difference between the mutual knowledge my parents and I had—of his nonexistence in this case—and the common knowledge we didn't have, not only knowing this fact but also knowing that the others knew, knowing that the others knew that we each knew, and so on.) My brother Paul, three and a half years my junior, was only a baby, so it wasn't him I was trying not to disillusion.

In any case, my qualitative "calculations" had proved to me that there were too many expectant kids around the world for Santa Claus to even come close to making his Christmas Eve rounds in time, even if he didn't stop for hot chocolate or bathroom visits, a pressing concern for me. This may sound like quite a pat memory for an author of a book titled *Innumeracy* to have,[1] but I do remember making rough "order-of-magnitude" estimates that showed Santa to be way overextended.

An oddly vivid memory of a similar idea dates from my days as a fifth grader. (It's a tiny bit jarring to even record that I had "days as a fifth grader.") The teacher was discussing some war, and a girl in the front row asked how the losing country survived if all its people died in the war. She cried when she spoke, and it was clear she believed that every single soldier as well as almost all other people in a losing country died. I remember wondering if, common sense aside, she'd ever seen the aftermath to World War II depicted in any of the countless cheesy war movies I used to watch every Saturday afternoon. I was very shy in elementary school and only quietly came to my smug judgment about her numerical naïveté. Besides I really liked her and, in a ten-year-old sort of way, thought her emotionality cute. (I plead guilty to having been a ten-year-old sexist.)

Before resuming the narrative, let me flash-forward for the next few paragraphs to note that these particular memories are not mere personal remembrances, but rather they connect up with later concerns of mine about numerical estimation and innumeracy. The sad not so cute reality is that most adults don't have a much keener appreciation for magnitudes than my fifth-grade classmate. To cite a once popular American TV show, I think it's fair to conclude that they are in this sense not smarter than a fifth grader, especially when the issue is numerical estimation.

Some easy examples from a course in quantitative literacy that I often teach illustrate the point as well as they do the persistence of my fifth-grade judgmental attitude. For example, if I say to students in my class or to adult acquaintances that I heard that a Rose Bowl quarterback once shook hands with almost everyone in the stands after a startling come-from-behind victory, they are rightfully dubious that this would ever happen. Still, very few will note that, like Santa's journey, it is next to numerically impossible.

Even if only half of the 100,000 fans or so at a game came down to shake hands and each handshake took, say, 4 seconds, that's 15 per minute, and 50,000 divided by 15 per minute is more than 3,300 minutes or about 7 8-hour days of handshaking. That would likely mean that the quarterback would never throw another pass.

Continuing with this perhaps jejune line of thought, I often get similar off-the-mark responses from students as well as educated neighbors if I claim to have read a headline claiming that experts fear housing costs (the total of rents and mortgages) will top $3 billion next year in the United States. They might respond by referring to the mortgage crisis, to greedy bankers, and the like, but they seldom will point out that the number is absurdly small—equaling about $10 annually per person for housing. And near total bafflement is the response to playful questions like "How fast does a human grow in miles per hour?" Alas, 62.38172548 percent of the time students and others take numbers as merely providing decoration, but not really imparting information.

Incidentally, a one-question diagnostic test for innumeracy is to ask someone to give very quickly and without resorting to a calculator the approximate average of these three numbers: 11 billion, 6 trillion, and 117 million. Only the numerate will answer 2 trillion. And a classic educational gambit assumes that the 4.5-billion-year history of the earth has been shrunk down to one year and asks you to determine how long before the end of the year various events occurred, say, the appearance of the most ancient religions as well as the appearance of you. Taking the first appearance of ancient religions to be very roughly 4,500 years ago and the first appearance of little old you, let's say, 45 years ago, the answers are only 30 seconds and .3 seconds, respectively, before midnight on December 31.

These little calculations are easily dismissible, but—and this is the point—I think they're examples of the same numerical obtuseness that afflicts many who think that a disproportionate share of American wealth goes to foreign aid, or that government earmarks for relatively paltry 50-million-dollar projects are the cause of the deficit, or that terrorism or Ebola, but not global warming, is a serious risk. And very few are aware that the two-trillion-dollar cost of the Iraq War is about 250 times the annual budgets of the National Science Foundation and the Environmental Protection Agency. Not unrelated is the mundane risk blindness of bicyclists I see driving down narrow, busy streets near my home in Center City, Philadelphia. Many don't wear a helmet, ride with no hands on the handlebars, text away on a cell phone, and listen to music through their earbuds. At the same time they might be very concerned about pesticide residue on the apple they are eating. Quite a stretch, I know, from beliefs about Santa Claus and bicyclists to foreign aid, war, and global warming, but this book contains at least as many general musings as personal remembrances, most of them, I hope, at least somewhat related.

To continue my story, at around this same time, the fifth grade or so, I began reading newspapers, an engaging introduction to which was provided by the *Milwaukee Journal*'s "Green Sheet." This four-page daily section was printed on green newsprint and was full of features that fascinated me. At the top was a saying by "Phil Osopher" that always contained some wonderfully puerile pun, a verbal category to which I'm still partial. There was also the "Ask Andy" column: science questions and tantalizingly brief answers. Phil and Andy became friends of mine. Were there a Twitter at the time and they had accounts, I would have been an avid follower and retweeter. And then there was an advice column by a woman with the unlikely name of Ione Quinby Griggs, who gave no-nonsense Midwestern counsel with which I often silently disagreed. Of course, I also read the sports pages and occasionally even checked the first section to see what was happening in the larger world.

Stimulated perhaps by Phil and Andy, I was captured at this young age by the idea of a kind of atomic materialism. I'd read that everything was composed of atoms, and I knew that atoms couldn't think, and so I "thought" this proved that humans couldn't think either. I was so pleased with this groundbreaking Epicurean idea (despite Phil, I didn't know the word yet) that I wrote it neatly on a piece of paper, folded it carefully, put it inside a small metal box, taped it securely, and buried it near the swing in our backyard where future generations of unthinking humans could appreciate my deep thoughts on this matter. I also remember speculating that just maybe there was a kid somewhere—in Russia, perhaps—who was as smart as I was. Pursuant to this I would scrawl "John is great" in secret places from closets to the attic, demonstrating either delusions of grandeur or simply youthful arrogance.

In any case the notion of emergent qualities, properties, and abilities didn't complicate my youthful certainty about these matters, and the dreary conclusion I came to that we couldn't really think was one I oddly found quite cheering. What I didn't find cheering was a recurring thought that some great new scientific discovery or philosophical insight would be announced, and I would find myself "seven brain cells short" of understanding it. It would be just beyond my personal complexity horizon. As a result of that irrational fear and of having read that alcohol kills brain cells, I resolved to be a lifelong teetotaler. My understanding of brains and conceptual breakthroughs has grown a little more sophisticated over the years, but the (almost) total teetotaling has persisted.

PEDAGOGY, VANQUISHING BLOWHARDS AND OPPONENTS, AND MONOPOLY

A bit later in elementary school I developed a very personal appreciation of mathematical certainty (as opposed to other sorts) that was germane to an adult concern of mine: mathematical pedagogy. Discussions of pedagogy and curricula too often tacitly assume there is a best way of imparting mathematical knowledge, igniting mathematical curiosity, and developing an appreciation for mathematics. There isn't. People's backgrounds, interests, and inclinations vary enormously and so should pedagogical techniques. As

I've described in *Innumeracy* and mentioned in the introduction here, my interest in mathematics proper was partially a consequence of an intense dislike for my elementary school mathematics teacher, whose real occupation, it seemed, was being a bullying martinet.

I was very interested in baseball as a kid. I loved playing the game and aspired to be a major-league shortstop. (My father played in college and professionally in the minor leagues.) Even now I vividly recall the highlights of my boyhood baseball career: a game-winning home run over a frenemy's backyard fence and a diving knee-skinning catch in center field on an asphalt playground. My two worse moments: being beaned at bat by the local fastball ace and racing in from center field only to miss catching the ball that sailed over my head.

I also became obsessed with baseball statistics and noted when I was ten or so that a relief pitcher for the Milwaukee Braves had an earned run average (ERA) of 135. (The arithmetic details are less important than the psychology of the story, but as I remember them, he had allowed 5 runs to score and had retired only one batter. Retiring one batter is equivalent to pitching 1/3 of an inning, 1/27 of a complete 9-inning game, and allowing 5 runs in 1/27 of a game translates into an ERA of 5/(1/27) or 135.)

Impressed by this extraordinarily bad ERA, I mentioned it diffidently to my teacher during a class discussion of sports. He looked pained and annoyed and sarcastically asked me to explain the fact to my class. Being quite shy, I did so with a quavering voice, a shaking hand, and a reddened face. (A strikeout in self-confidence.) When I finished, he almost bellowed that I was confused and wrong and that I should sit down. An overweight coach and gym teacher with a bulbous nose, he asserted that ERAs could never be higher than 27, the number of outs in a complete game. For good measure he cackled derisively.

Later that season the *Milwaukee Journal* published the averages of all the Braves players, and since this pitcher hadn't pitched again, his ERA was 135, as I had calculated. I remember thinking then of mathematics as a kind of omnipotent protector. I was small and quiet and he was large and loud, but I was right and I could show him. This thought and the sense of power it instilled in me was thrilling. So, still smarting from my earlier humiliation, I brought in the newspaper and showed it to him. He gave me a dirty look and again told me to sit down. His idea of good education apparently was to make sure everyone remained seated. I did sit down, but this time with a slight smile on my face. We both knew I was right and he was wrong. Perhaps not surprisingly, the story still evokes the same emotions in me that it did decades ago.

So, is what this teacher did good pedagogy? Of course not, and happily I benefited from many more knowledgeable, supportive, and nondirective teachers and a variety of pedagogical approaches. Nonetheless, this particular teacher did give me a potent reason to study mathematics that I think is underrated. Show kids that with it and logic, a few facts, and a bit of psychology you can vanquish blowhards no matter your age or size. Not only that, but you can sometimes expose nonsensical claims as well. For many students this may be a much better selling point than being able to solve mixture problems or using trigonometry to estimate the height of a flagpole from across a river.

Not unrelated is another bit of mathematical pedagogy that was of benefit to me early on: board games, Monopoly in particular. In this game players roll dice to move around a board lined with properties (as well as railroads and utilities), which they can purchase with the game's money, develop by purchasing houses and hotels to occupy them, and from which they can derive rents from the other players who happen to land on these board spaces. The point of the game is the lovely one of driving one's opponents into bankruptcy. Like the best teaching, it's invisible.

For example, to determine how likely I was to land on Boardwalk or Jail I needed to figure out the probability of the various outcomes when rolling a pair of dice. Obtaining a sum of 7, I realized, was the most likely outcome, arising in 6 out of the 36 possible outcomes—(6,1), (5,2), (4,3), (3,4), (2,5), and (1,6)—whereas 2 and 12 were the least likely, each arising in only 1 of the 36 possible outcomes—(1,1) or (6,6). As anyone who has ever played the game knows empirically, Jail is the square on which players spend the most time. Thus the orange properties are good ones because they're relatively cheap and visited often by those leaving Jail.

My childhood Monopoly discoveries and competitiveness are not the point here. Rather, it is that implicit in the game, whether young players realize this or not, are a number of important mathematical ideas that one gradually absorbs while playing the game, in my case for untold hours on boring summer afternoons. Among these ideas are probability, expected value (average payouts per owned property), even Markov chains, which explain which squares are most likely to be landed upon. (The latter are systems like Monopoly that transition from one state—a player's place on the board, say—to another, the next state depending only on the present one.)

The lessons imparted by the journey around the Monopoly board are in some obvious ways relevant to the journey through life. Being cognizant of probabilities; being sensitive to possible playoffs, risks, and rewards; and being aware of long-term trends are all useful life skills.

The game, not to mention the vagaries of life, also allows for ad hoc and difficult-to-quantify changes to the rules. One such rule I remember is that if an adult enters the room when a player is rolling the dice, that player must pay a fine of $2,000. We soon abandoned this rule, however. It led to arguments when a player who was behind began suspiciously and loudly coughing when it was the leading player's turn to roll the dice. Related problems arose with the rule that allowed an occasional looting of the Monopoly bank.

Unfortunately board games are a bit passé today, but many (not all) of the same lessons and other quite new ones can be had via video games. In fact, mathematics professor Keith Devlin has advocated the use of appropriately chosen games as vehicles for imparting mathematical ideas to middle school players. Note that I used word *players* rather than *students* intentionally. The games generally are chases or fights in which the players must solve puzzles and devise strategies to best their opponents.

OF MOTHERS AND COLLECTING BASEBALL CARDS

An almost canonical story that straddles memoir and (in my case) math is about mothers who think it is their duty to throw away arguably childish but still valued possessions when one reaches a certain age. My experience confirms the cliché and illustrates a nice bit of mathematics. As a young boy I was an avid collector of baseball cards, and for a couple of years in the late '50s I managed to collect the complete set of cards that usually came in packs of five along with a piece of pink bubble gum that I loved but would now characterize as revoltingly sugary. I remember how long it took to obtain the last two or three cards of the collection that I hadn't yet obtained. If memory serves, the last holdout was that of Charlie Grimm, who at the time was manager of either the Chicago Cubs or Milwaukee Braves, the acquisition of whose card required the purchase of hundreds of cards, all except one of which were doubles, triples, quadruples of cards I already had. After completing the set I put them in a little box, labeled them, and some years later went off to college. I've already revealed the punch line. Searching my drawers a while later, I discovered the cards missing and learned that my mother had thrown them out thinking I'd outgrown them. Since I rarely even looked at them, this was a reasonable but unsatisfying assumption. At the very least, had I held onto them into the age of eBay, I might have sold them for a nontrivial amount of money.

The silver lining to this prosaic story is that it got me thinking about a notion I vaguely understood but would soon learn mathematicians called the expected, or average, value of a quantity. In particular I wondered how many baseball cards on average one would need to buy before one obtained the complete set of, say, 400 different cards (making the dubious assumption that the companies printed the same number of each card). Probability theory[2] says that to obtain all x cards, one would need to buy approximately x × ln(x) or x times the natural logarithm of x cards. (The natural logarithm is like the notion studied in high school algebra but with the base e, which will make several later appearances herein, rather than with the base 10.) This very nice result, which involves adding up a bunch of so-called geometric random variables, says that one would need to purchase approximately 2,400 cards, on average, to obtain the complete set of 400. No wonder I had more than my share of cavities in those days.

The same analysis applies to any set of items you obtain randomly and that have roughly equal probabilities of occurring. Stretching the conditions a little, you might try to say how long it would take you to achieve this or that collection of everyday adult experiences—stubbing your toe, having someone skip in front of you in line, losing some little item. Note that the analysis above doesn't apply to significant experiences you aim for, such as a so-called bucket list of places you'd like to visit before you die, since these experiences do not occur randomly. Even this situation, however, is subject to a similar though more complicated analysis. We can consider different probabilities of attaining each element on your bucket list, for example.

(A little explanation of the collection problem that is eminently skippable: let's check that, on average, you would need to roll a die 14.7 times before "collecting" all six numbers on it [or all six action figures in some brand of children's cereal]. The first time you roll the die you will, of course, get one of the six numbers. After this happens, the probability of obtaining a different number on the next roll is 5/6, and so it will require a bit more than one roll to get the second number. It can be shown that on average it will require 1/(5/6) or 6/5 rolls to get the second number. After this happens, the probability of obtaining a third different number on the next roll is 4/6, and so it will also require more than one roll to get the third number. It can be shown that on average it will require 1/(4/6) or 6/4 rolls to get the third different number. We continue in this way. After obtaining 3 different numbers, the probability of obtaining a fourth different number is 3/6, and so on average it will require 1/(3/6) or 6/3 rolls to get the fourth different number. Likewise, to get the fifth and sixth different numbers requires 6/2 and 6/1 rolls, respectively. Summing 1, 6/5, 6/4, 6/3, 6/2, and 6/1 gives us 14.7 rolls, on average, to "collect" all six numbers.)

My baseball cards were thrown out, but the math stayed with me. It was much more valuable anyway.

A FURTHER NOTE ON MATH, HUMOR, AND MY EDUCATION

In high school I was what would today be called a nerd, but I was considered by some classmates a cool instance of one. I both smile and cringe to recall how cool (such an uncool word now) I thought I was pantomiming the Everly Brothers and Elvis Presley with my imaginary guitar and swirling pompadour. (Selection bias on my part induced me to internalize the statements of these friendly classmates.) I dated hardly at all, and pornography—*Playboy* magazine was the closest approximation to such—was nowhere near as available as today; there was precious little real or virtual sex at Milwaukee's Washington High School. I did read a lot, however, and in the summer between my sophomore and junior years of high school I spent every morning under the tree in my grandmother's front yard in Denver attempting to review and copiously supplement everything I'd studied up to that time, including old textbooks, novels, biographies, essays. Montaigne was a particular favorite, as was the math columnist Martin Gardner and even the tabloid *Rocky Mountain News*. I remember thinking that I had established a solid foundation for future academic/intellectual pursuits and thus that

that summer was somewhat pivotal. In retrospect those days, the mornings in particular, seem idyllic, safe, and anxiety-free, but as with all such memories I wonder if they really were as remembered.

High school is often described as a particularly turbulent time in a person's life, but this wasn't the case with me. I recall having three desires at the time: to learn, to get away from home, and to have sex. As I mentioned, I did manage to learn a lot before I went away to college.

At one time or another in high school or as an undergraduate at the University of Wisconsin in Madison I majored or contemplated majoring in classics, English, philosophy, physics, and, of course, mathematics. I loved Latin and reading Caesar and Virgil and diagramming the Latin sentences in them, even though the ablative absolute didn't lend itself to modern conventions. My interest in classics, however, waned over time, as did others.

Despite the brief separations and flings with the above disciplines and other topics, I gradually became more deeply enthralled with the beauty, elegance, and power of mathematics. One example that struck me at this time: randomly pick a number between 0 and 1,000, say, 356.174, and then pick another, say, 401.231, and add it to the first to get 757.405, and keep doing this until the sum exceeds 1,000. On average, how many random numbers need you pick for their sum to exceed 1,000? The answer turns out to be e, which is approximately 2.71828, a non-repeating decimal that is not the root of any algebraic equation; is, as noted, the base of the natural logarithm; lies at the foundation of compound interest, mortgages, annuities, and modern finance in general; and is as pervasive a celebrity in the community of numbers as its better known cousin pi.

My fascination with e has not waned. Recently I described in an article on ABCNews.com how the number e lurks even in the night sky.[3] To see this, imagine dividing some square portion of the night sky into a very large number, N, of smaller squares. Find the N brightest stars in this large portion of the sky and count how many of the N smaller squares contain none of these N brightest squares.

Call this number U. (I'm assuming that the stars are distributed randomly, so by chance some of the smaller squares will contain one or more of the brightest stars, others none at all.) I won't do so here, but it's not difficult to show that N/U is very close to the very same number e and approaches it more and more closely as N gets large. (A different analysis shows that pi also resides quite naturally in the night sky.)

Easily stated open problems also fascinated me (and still do). An example that incidentally also intrigued my son when he was in college is the so-called Collatz (3×+1) conjecture: start with any whole number. If it's odd, multiply it by 3 and add 1, but if it's even, divide it by 2. Follow the same rule with the resulting number and continue doing so with the succeeding numbers. The conjecture is that the sequence of numbers generated in this way will always end 4, 2, 1, 4, 2, 1, . . . For example, assume you started with 23. The sequence would then be 23, 70, 35, 106, 53, 160, 80, 40, 20, 10, 5, 16, 8, 4, 2, 1, 4, 2, 1. . . . The conjecture has been checked with extremely large numbers, but it has never been proved.[4]

Why these unexpected and amazing connections that transcend all divisions of nationality, culture, gender, class, and time? Clearly their truth is not affected in the least by the personal characteristics of the people who discover/invent them. An exemplar of this universality is Srinivasa Ramanujan, the untutored Indian genius who contacted the British mathematician G. H. Hardy about his astonishing mathematical insights and theorems and traveled to England to develop them but died there at an early age. In Robert Kanigel's biography of him, which I reviewed for the *New York Times*, there is a very moving vignette of the boy Ramanujan alone in the shadows of a Hindu temple, his elbow dirty from repeatedly erasing his chalk slate, as he cogitated and calculated away.[5]

The universal beauty of mathematics seemed (and still seems) ethereal to me, its elegance was mesmerizing, and its reach appeared limitless. I remember wondering about the certainty of these sometimes mysterious mathematical truths. Might it derive from or be reduced to the laws of logic, and might mathematical statements just be circuitous ways of invoking the law of the excluded middle, "A or not A"? Or might mathematical truth be simply a matter of convention more complicated, but not any more inexplicable, than the fact that 27 cubic feet equal 1 cubic yard or 16 ounces equal 1 pound? Might it simply be a discipline governed by rules, much as the movement of pieces on a chessboard is governed by the rules of chess?

Or might mathematical truth be a reflection of the activities of numbers and figures cavorting in some sort of Platonic heaven? Is this where unfathomably large numbers—like 52! (52 × 51 × 50 × 49 . . . × 2 × 1), almost equal to 1 followed by 68 zeroes, the number of ways to order a deck of cards—live and play? (This latter number is so mind-bogglingly huge that the order of any well-shuffled deck might rightfully be called a miracle since the probability of its occurring is almost impossibly, minusculely minuscule.)

Why was mathematics so useful? These questions, the actual theorems I was studying, and the ever-growing list of practical applications of mathematics both vexed and captivated me.

Another contributor to my enthrallment, I must admit, was the much earthier realm of humor. I'd always appreciated literal interpretations of figurative phrases (such as "this is only a fraction of what you'll pay elsewhere," where the price is 5/3 of that elsewhere), self-reference, unusual juxtapositions and permutations, logical paradoxes, and incongruities of one sort or another, all elements of humor and, what is less well known, mathematics. In fact, as I've written elsewhere, ingenuity and cleverness are hallmarks of both humor and mathematics as is a Spartan economy of expression. Long-windedness is as antithetical to pure mathematics as it usually is to good humor. At the risk of being long-winded myself, I'll note that the beauty of a mathematical proof often depends on its elegance and brevity, qualities I prized even as an arrogantly dismissive teenager.

My cast of mind predisposed me to the study of logic and mathematics, which in turn furthered my natural attraction to such humor. I saw Groucho Marx's *Duck Soup* several times and even now remember exchanges from the movie. Minister of war: That's the last straw! I resign. I wash my hands of the whole business. Firefly (Groucho): A good idea. You can wash your neck too. And, as mentioned, I still like puerile jokes: Teacher: Johnny, name two pronouns. Johnny: Who? Me? And I hate unmatched (parentheses. Much later after moving to Philadelphia I even tried stand-up comedy, albeit rather unsuccessfully, and learned that my taste in "logical humor" was not widely shared. A corollary is that three minutes is a really long time with material similar to jokes that generally only mathematicians or logicians seem to like: "Spell 'Henry.' That's easy, HENRY. No, it's HEN3RY. The 3 is silent" or "I'd give my right arm to be ambidextrous."

Although I dated some in college, where humor of a warmer sort was more of an asset than math, I spent most of my little free time with some good friends, a couple of whom I retain to this day, and continued my studious ways. The story of my collegiate years is devoid of any overt drama, although becoming acquainted with the Banach–Tarski theorem (math joke: the Banach–Tarski–Banach–Tarski theorem; the explanation is that theoretically a sphere the size of a tennis ball can be decomposed and reconstructed into a sphere the size of a basketball or even one the size of the sun)[6] and Kurt Gödel's incompleteness theorems on the inevitable limitations of formal mathematical systems were exciting enough for me. I never did any drugs except for one silly attempt to smoke banana peels, which seemed like an appealingly natural way to get high. All in all, I loved Madison, and the names Rathskeller, Lake Mendota, Van Vleck, Picnic Point, and Bascom Hill still evoke pleasant memories.

People sometimes assume a conflict between enjoyment and hard work. If it's not pushed too vigorously, I think such a trade-off sometimes does exist. On the other hand, hard work, especially on a project, subject, or endeavor that one loves, is itself enjoyable. In any case, moving to the University of Washington in Seattle for my master's degree resulted in more of the same, the scholarly idyll interrupted only by the specter of the draft and the Vietnam War. Rather than face the prospect of fighting in a war I very much opposed, I decided to temporarily abandon my PhD studies and enter the Peace Corps in Kenya, where I taught math at Kakamega Secondary School and became acquainted with a world quite different than that of a graduate school in mathematics. (With more punch than sense, the innumerate might say 360 degrees different). I later returned to Wisconsin for my PhD in mathematics.

Chapter 2

BIAS, BIOGRAPHY, AND WHY WE'RE ALL A BIT FAR-OUT AND BIZARRE

BIAS AND MINDSETS, STATISTICS AND BIOGRAPHY

I had a number of professors at University of Wisconsin who, I was warned, were horrible; some were said to mumble, others were given to excessive abstraction, a few constantly digressed. Many just lectured. When I enrolled in their courses despite these warnings, I was often surprised. Turns out I usually liked abstract, mumbling digressers. And I much preferred hearing a lecture from someone knowledgeable than listening to fellow students getting together in study groups and giving their usually uninformed perspectives on the topic. Contrariwise, I was often disappointed by those "fun" professors whom most deemed wonderful. The same phenomenon holds for people about whom I've heard only bad (or good) assessments that I find to be baseless after meeting the people in question. I sometimes still get annoyed at my own trustingness.

This, of course, is not a particularly unique realization. Everyone has experienced variants of it. One doesn't have to be too statistically savvy to know that comments and surveys that derive from only a dozen or so people are not very reliable. Neither is it arcane mathematical knowledge that self-selected samples are not very likely to be representative of the population at large. A small self-selected sample of people responding to a television "poll" about more stringent gun control, for example, may very well arouse a disproportionate number of passionate NRA members and significantly skew the results.

Another example of wayward statistics in academe: Like many universities, mine requires that students take a core survey course in mathematics if they're not going on in the subject. Passing the course requires that a student's grade be at least a C-. Suspecting that the number of C-'s would be much larger than the number of D+'s because of how crucial this small difference is, I decided to examine the number of C-'s and D+'s given in this course over the years for which I could find the records.

Sure enough, I found that approximately eight hundred C-'s and 100 D+'s were awarded. Someone might point out that the number of D+'s should be lower than the number of C-'s simply as a result of the normal bell-shaped distribution with an average of C or so. This, however, cannot be the explanation, since the drop-off was so precipitous, eight times as many C-'s as D+'s. (The four hundred or so plain D's and roughly seven hundred F's given out during this period indicate that general grade inflation was not the issue.) The reason was probably that, at the crucial cutoff between C- and D+, the faculty were likely to give students a bit of a break. Assigning grades is not a cut-and-dried activity, and many professors seemed to have given students the benefit of a doubt in these close calls rather than being blindly bound to rigid grading that is inevitably somewhat arbitrary.

To vary the examples a bit, consider the museum guard who claimed that a dinosaur on exhibit was 70,000,009 years old. Asked how he knew that, he said that he had been told it was 70,000,000 years old when he'd been hired nine years before. The precision would be laughable, but shouldn't we find it almost as laughable when someone claims to be relating someone else's verbatim (precise) conversations as well as their dates, locations, and contexts?

Why are such elementary understandings and explanations seldom invoked in our reading of personal profiles or full-scale biographies? Biographers (and, of course, autobiographers) select themselves in part because they resonate in one way or another with the subject. They may interview many people about their subject, but even their choice of interviewees is likely to be influenced by their biases. So are the questions they ask, rephrasing them over and over if they don't get the answer they want. Pick almost any potential biographical subject and ask ten people who know him or her what they think of the person, and the responses will certainly be quite varied. Picking only eyewitnesses to events in a subject's life certainly doesn't guarantee truth either. Recall the trial lawyer's quip "The only thing worse than one eyewitness is two eyewitnesses."

Here's a simple thought experiment: consider just yourself as a subject and think of who in your background would write the most scathing biography of you, who would write the most sympathetic account (C- rather than D+ on the scale of a whole life), and who would write the most clueless one. Or, if it doesn't hurt too much, imagine a biography of Stephen Hawking written by Kim Kardashian and one of her written by him. Or come up with your own incongruous pairs of reciprocal biographers. The difficulty of adopting the perspective of a biographical subject's perspective is suggested by the story of two strangers walking on opposite sides of a river. One of them yells across, "How do I get to the other side of the river?" The second one answers, "You are on the other side of the river."

Any story of adultery, to cite one last example, will read quite differently depending on which of four natural biographers write it: the injured spouse, the wandering spouse, the outside lover, or a "neutral" observer. It's interesting to imagine *Madame Bovary* from Charles Bovary's point of view. The fake 2007 newspaper headline in the *Onion*, a satirical magazine, makes the same point: "Majority of Parents Abuse Children, Children Report."[1] Despite these obvious concerns, most people assume biographies or magazine profiles or even informal spoken descriptions of a person are more or less accurate. You might have inferred that this annoys me.

Phrasing the issue in the jargon of mathematical logic, I note that "is a biography of X" is a so-called unary predicate, and it would be preferable if it were replaced by the binary predicate "is a biography of X written by Y." Perhaps it's even more prudent to consider ternary predicates: "is a biography of X written by Y at time Z." An "autobiography of X," by contrast, really is a unary predicate unless you happen to be a schizophrenic. Comedian Steven Wright's quip that he was writing an unauthorized autobiography also comes to mind.[2]

Ideally nasty biographies should give a rough measure of the ratio of research undertaken to secrets uncovered. Laudatory ones as well should provide a ratio of the time taken to the positive tidbits found. Autobiographies should be scrutinized for any traces of the Lake Wobegon effect, whereby the author and everyone closely associated with him or her is above average. The problem, of course, is that if one looks hard enough, one will likely find what one is looking for. We're all subject to confirmation bias, the tendency to look largely for confirmation of our hunches and beliefs and rarely for disconfirmation, but perhaps few more so than biographers, who are often either in thrall to their subjects or else detest them.

Whether reading a life story or just listening to a neighbor, we should be aware that very many of our beliefs and attitudes are likely to be a consequence of probabilistic misunderstanding and statistical failings, bad sampling in particular. Most people, for example, become somewhat more reclusive when depressed or otherwise behaving "abnormally," so these behaviors will be under-sampled and thus likely will play a smaller role in their biographies than they do in their lives. Likewise, successful people (as well as their biographers) will tend to see a strong connection between their personal qualities and their success even if they self-effacingly say how lucky they've been; conversely, less successful people will tend to see a weak connection. Neither viewpoint is statistically robust.

Regarding confirmation bias and success, consider an experiment I described in one of my columns for ABCNews.com in which subjects were told of two firemen—one successful, one not. Half the subjects were told that the successful fireman was a risk-taker and that the unsuccessful one was not. The other half of the subjects were told that the successful one was not a risk-taker and that the unsuccessful one was. Afterward, they were informed that the firemen did not exist and that the experimenters had simply invented them.[3]

Amazingly, the subjects continued to be strongly influenced by whatever explanatory stories they had concocted for themselves. If they had been told that the risk-taking fireman was successful, they thought that prospective firemen should be chosen for their willingness to take risks; if not, then not. If asked to account for the connection between risk taking or its absence and successful firefighting, the members of each group concocted a confirmatory explanation consistent with the imaginary story originally told them.[4] What holds for firemen no doubt also holds for statesmen (as well their biographers).

One other quite significant, albeit underappreciated, point about statistics in the presentation of biographies bears repeating. As I mentioned in the introduction, an important aspect of a story is that there is a tendency to suspend disbelief while seeing, reading, or hearing it so as to not spoil its enjoyment. "Let's pretend there really is a monster like this." Loosely paraphrased in statistical parlance, this means that one risks a so-called Type I error (a false-positive); that is, saying an important incident or phenomenon occurred that really did not. This is not the situation in statistical or scientific contexts where one typically suspends belief so as to not be fooled. "How do we know that?" In statistical parlance, one risks making a so-called Type II error (a false-negative); that is, saying that an important incident or phenomenon that really did occur did not.

Both biographers (storytellers generally) and statisticians (scientists generally) wish to avoid both sorts of errors, but biographers are a bit more careful about not ruling things out, scientists a bit more careful before admitting them. What exists for the two sorts, their ontology, is different; for storytellers it's generally more baroque, for scientists more bare-bones. To parody this difference in mindset, we might say that out of five crises, storytellers predict seven of them, and scientists predict three of them—hyperbolic versus hypobolic, which should be a word.

Type I and Type II errors are part of a complex of notions surrounding Bayes' theorem, an extremely seminal result in probability theory that tells us how to update our probabilities in the light of new evidence. For example, if a fair coin (H-T) and a two-headed coin (H-H) are on a table and we choose one of them, the probability we choose the fair coin is 1/2. But if after we've chosen a coin, we flip it three times and obtain three consecutive heads, Bayes' theorem tells us the probability that we chose the fair coin shrinks to 1/9.

This is, of course, much harder to do with more nebulous stories and biographies, but do biographers make much of an attempt to indicate how they come to their initial evaluations of a subject or how they've changed their minds about him or her in the light of new documentary evidence? I doubt it. Certainly excusable, but why make so little use of possibly relevant mathematical and scientific tools?

One proto-Bayesian, the empiricist Scottish philosopher David Hume, underlined the importance of considering the probability of supporting evidence when he questioned the authority of religious hearsay: one shouldn't trust the supposed evidence for a miracle, he argued, unless it would be even more miraculous if the report were untrue. In ancient times, biographies of saints and kings were replete with miracles. Contemporary biographies are devoid of miracles but still contain too many exploits and adventures that seem considerably less likely than their nonoccurrence. It's the same impulse, but attenuated.

Rare events such as accidents, hurricanes, and lottery oddities are much better described by the often amazingly accurate Poisson statistical distribution. This gives us the probability of a given number of events during a given time interval if the events occur at a known average rate independent of the elapsed time since the last event.

DESPITE NORMAL APPEARANCES, WE'RE ALL STRANGE

These considerations and perhaps even the fairly anodyne personal anecdotes related above suggest an idea relevant to biographies and the often simplistic way we think of others. Most of us have had the experience of knowing someone superficially and thinking him or her to be quite normal, exemplary even—wonderful family, productive career, public-spirited, and so on. Then, if and when you get closer, you discover that the person is quite bizarre in some ways—not necessarily bad ways, but just very odd ways.

Think perhaps of the television series *Breaking Bad* and the brilliant but poor chemistry teacher and family man dying of cancer who decides to use his expertise to become a manufacturer and distributor of crystal meth. I once was acquainted with someone who appeared to be completely conventional, totally banal, and utterly unimaginative. Then a friend of mine saw him on several occasions gluing five-dollar bills to the sidewalk and then, having retreated a distance from them, giggling at people trying to scrape them off. This is not a confessional book, but, despite being generally quite honest, I do remember a few times sticking a Chuckles candy into a *Wisconsin State Journal* when I was in college and paying only for the latter. These incongruities bring to mind the quip: There are two kinds of people in the world—those who are very strange and those whom you don't know well.

If by "very strange" or bizarre people, you mean people who, along at least some measurable dimensions, are statistically way off the charts, then this is almost certainly true. "Dimension" can be something geometrical, but needn't be. Think, for example, of dating services that advertise that they check prospective couples for compatibility along dozens of possible dimensions—aspects of personality, unusual obsessions, fears, hobbies, family backgrounds, politics, etcetera. They might consider fifty or more such dimensions.

Or consider people as consumers whose tastes differ along many more dimensions. One can ask whether they prefer one brand of fig bar to another or, more generally, one kind of product to another. We can inquire as well how much they like their fig bar or, more generally, about the intensity of their preferences. Of course, we can also include dimensions of sexual variety, which I'll leave to readers' experiences and imaginations (or to pornhub.com for the unimaginative).

A simple invocation of what's called the multiplication principle in combinatorics is sufficient to demonstrate the number of sexual varieties. Biologically, people can be male, female, or intersex—3 possibilities. They can be attracted to males, females, both, or neither—4 possibilities. And independent of these categories their gender identity may be male, female, androgynous—3 possibilities. Multiplying 3 by 4 by 3 yields 36 different varieties, all A-OK. Further divisions (involving the ages of the people to whom one is attracted as well as various paraphilias) would, of course, increase this number significantly. More generally, products of the numbers of possibilities grow rapidly and lead to huge total numbers of possibilities, not just for sexual variety but also along countless other (collections of) dimensions.

If we define people solely in terms of numbers along these various dimensions, that is, reductively as a collection of atomic traits and as nothing more coherent, then we can more easily understand geometrically why we are all quite strange and far-out. Why exactly?

Toward a geometric understanding of this, imagine a straight line 10 inches long, along which people can be measured on some dimension of interest. Let's consider the parts of the line within 1/2 inch of either end and call it the extreme part of the line. The normal part is the 9-inch middle section, which constitutes 90 percent of the line.

Now consider a square 10 inches on a side, along both of whose dimensions people can be measured. Consider the part within 1/2 inch of a side of this square and call this border area the extreme part of the square. The normal part of the square is the middle section, which constitutes 81.0 percent of the square. This can be seen by noting that the whole square is 10^2 or 100 square inches, and the normal part constitutes only 9^2 or 81 square inches. If you like circles better than squares, realize that the interior of a 10-inch pizza with a half 1/2-inch crust all around is likewise only 81 percent of the area of the whole pizza. A 1-inch crust leaves only 64 percent for the interior; $8^2/10^2 = .64$.

Or consider spherical potatoes and my Thanksgiving epiphany. During turkey day preparations recently, I was peeling potatoes in my usual slapdash way and wondered how much of the potato I was wasting. For simplicity's sake I assumed the potatoes were spherical and about 10 centimeters (4 inches) in diameter and that my lazy, sloppy peeling removed about 1 full centimeter from the surface. The peeled potato was thus 8 centimeters in diameter, and the ratio of its volume to that of the unpeeled potato was 512/1,000 (which is $8^3/10^3$). I was wasting almost half the potato! A disproportionate fraction of its volume was in its periphery.

Next up, let's picture a cube 10 inches on a side, along all three of whose dimensions people can be measured. Consider the part within 1/2 inch of an outside face of this cube and call it the extreme part of the cube. The normal part of the cube is the middle section, which constitutes 72.9 percent of the cube. This can be seen by noting that the whole cube is 10^3 or 1,000 cubic inches, and the normal part constitutes only 9^3 or 729 cubic inches. Returning to Italian food, I note that a similar analysis applies to spherical meatballs.

Although it can't be pictured as easily, the same idea makes sense with higher-dimensional hypercubes. Imagine, for example, a four-dimensional hypercube 10 inches on a side, along all four of whose dimensions people can be measured. (Pick your favorite four.) Consider the part within 1/2 inch of the outside of this hypercube and call it the extreme part of the hypercube. The normal part of the hypercube is the middle section, which constitutes 65.6 percent of the hypercube. This can again be seen by noting that the whole hypercube is 10^4 or 10,000 cubic inches, and the normal part constitutes only 9^4 or 6,561 cubic inches.

Note that as the number of dimensions increase, the normal part of the hypercube constitutes a smaller and smaller part of the volume of the hypercube in question. We can continue this game and consider not 4 but 50 dimensions along which people can be measured and perform the same sort of calculation. If we do, we'll find that that the interior or normal part of the resulting hypercube constitutes only about 1/2 of 1 percent of the volume of the hypercube. For 100 dimensions, the interior or normal part shrinks to only .0027 percent of the total volume!

And this says what exactly about our abnormality, our strangeness?

Note that most points in the hypercube will be extreme along at least some dimensions and hence will be at the extreme edges of the hypercube. In this same sense (as well as in others), most people live on the extreme, abnormal edges of the human multi-dimensional hypercube. Vanishingly few of us live in the moderate, normal interior part of the hypercube. Remember that we are defining people reductively as large sets of numbers ranking us along many different dimensions (that is, as points in the hypercube). For the purposes of this exercise each of us can be considered an atomistic collection of preferences, ranging from Prada shoes and Smucker's jelly to investment positions and voting tendencies. (Note, too, that there are surely many more than 50 or 100 dimensions of various sorts along which people can differ and that we can define an extreme score along any given dimension to be one that's even more extreme—as, say, the top and bottom .5 percent rather than the top and bottom 5 percent of the given dimension.)

Once again, if each of us has a score along each of the very many dimensions in a hypercube, then almost all of us will find ourselves to be a point along the edges of the hypercube; that is, an extreme, abnormal point. Nobody except the hopelessly boring and banal live in the moderate, normal interior of the human hypercube.

The same sort of argument can be made in probabilistic terms rather than geometric ones and can also employ the so-called normal distribution (*normal* is unfortunately a much overused word in mathematics) rather than the uniform, flat distribution I assumed above. Other statistical distributions that lead to even more extreme oddities also exist. Consider an adoptable trait that is slightly unusual or abnormal but only in a statistical sense—say, for example, extended body tattoos. When people consider adopting it, they're influenced by others they know or hear about who have done so, and thus they are marginally more disposed to develop the trait themselves. That creates a cascading effect that results in an "abnormal-get-more-abnormal" sort of phenomenon, as the trait and (some of) the people holding it become more and more extreme. This is one reason so-called power law distributions (where one quantity varies as the power of another) are so common in social situations.

I like this little dimensional exercise since it reminds me of the classic book *Flatland*. The story employs a fictional two-dimensional world of squares and polygons (with a visitor from the third dimension) to make social observations about Victorian culture.[5] A much paler, much reduced bit of social commentary might be squeezed out of the above segment as well, something along the lines of increasing tendency to define ourselves reductively as a mere collection of traits, interests, consumer preferences, and so on.

In any case, given this suggestive little model of multidimensional human beings, it's not really very surprising that there are two kinds of people in the world—those who are very strange and those whom we don't know very well. Once again, almost all of us live in the outer shell of a multidimensional hypercube (or hypersphere or hypermeatball), whose interior is largely devoid of other humans.

None of this, however, contradicts the observation that there is a broad area of similarity among all people that is described by German mathematician Carl Friedrich Gauss's bell-shaped normal distribution. Whenever we consider traits that depend on many independent factors, the pervasively applicable central limit theorem of statistics tells us these traits will be normally distributed. Along most dimensions people are fairly nondescript and ordinary, even though almost all of us are quite extreme and extraordinary with respect to a good number of other dimensions.

MISAPPLICATIONS OF MATHEMATICS TO EVERYDAY LIFE— A CAVEAT

An acquaintance of mine in graduate school would begin what he considered to be a conversation with "Let X be a completely normed Banach space" (an abstract vector space satisfying certain conditions) and would then proceed to the statement of a theorem and its proof, all the while avoiding any eye contact whatsoever. He also tended to see almost everything as a Banach space. Needless to say, his equivalences and analogies to real-life matters were strained at best, more often totally cockamamie. Perhaps, to an extent, this is true of me.

His myopic view of the world as a collection of different sorts of Banach spaces raises a caveat about the application of mathematics to everyday life or, as in biographies, to everyday lives. The caveat concerns the risk of scientism and, although it's not a real word, mathism. I'm referring to an exaggerated trust in the notions and methods of science and mathematics, especially outside of their usual and warranted domains of application. It takes a certain finesse to decide whether a nonstandard application, model, or analogy is helpful and clarifying or misleading and silly. People in general, not just those of my acquaintance, tend to wrap themselves in the indubitability of mathematics and pretend their particular use of it is similarly indubitable, but as statistician George Box once observed, "All models are wrong; some models are useful." He also observed, "Statisticians, like artists, have the bad habit of falling in love with their models."[6]

Forget Banach spaces, however. Here's an easy example. It's a truism that two plus two equals four, but does it always? Not really. If one takes two cups of water and adds it to two cups of popcorn, one will probably get around three cups of soggy popcorn, not four. Even this most trivial fact of mathematics can be misapplied.

Or consider the tribe of bear hunters that died off shortly after mastering vector analysis, the study of quantities that have both a magnitude and a direction. Before hearing of this branch of mathematics, whenever they spotted a bear to the northwest, they would simply shoot their arrow in its direction and kill the bear. Afterward, however, whenever they spotted a bear to the northwest, they would shoot one arrow to the north and one arrow to the west, as vector analysis stipulates that we should add vectors, and the bear would get away. Again, we have a misapplication of a powerful bit of mathematics.

In these two cases it's clear that the applications are nonsensical, but what if the mathematics is more esoteric, for example, some complicated regression or invocation of an unusual dynamical law or the use of the wrong geometrical or algebraic structure or the unwarranted use of the Black–Scholes formula for evaluating stock options or even the obsessive references to Banach spaces?

Related to the issue of using inappropriate mathematical models (from simple addition to vector analysis to more esoteric matters) in an attempt to clarify some phenomenon is doing the reverse: looking at a well-understood phenomenon and trying to retrofit mathematical models that are broadly consistent with it. This is sometimes done with the hope that the model will shed some oblique light on the phenomenon. A recent book by Michael Chwe, for example, tries to make the argument that in her novels Jane Austen intuitively understood and was using various notions from game theory, a mathematical theory of strategic decision making involving notions such as expected value and zero-sum contests. The book also suggests these mathematical notions give us a slightly different slant on Austen's mindset.[7] Likewise, mathematicians have analyzed the dynamics of the love affair between Scarlett O'Hara and Rhett Butler in *Gone with the Wind* utilizing a pair of differential equations, one for her and one for him. This reverse process is also a perilous but often interesting employment of mathematical ideas as the above two examples are.

In any case, given this book's topic of mathematical musings on biography and the temptation it affords, I've tried my best to avoid scientism and mathism here and have issued this warning since I'm not certain that I have completely succeeded. That said, I reiterate my belief that there is great value in trying to apply mathematical ideas in novel ways to almost any discipline (or nondiscipline). I should also add that the distinction between pure and applied math is largely bogus. The attempt to model any phenomenon from bank transfers to bridge crossings to (perhaps) biographical ruminations can very well lead to a bit of pure mathematics. Likewise even the purest of mathematicians such as G. H. Hardy and Bernhard Riemann often do work that is later seen to be of seminal importance in various scientific fields, general relativity in Riemann's case. Despite the nonexistent distinction between pure and applied mathematics, Hardy in his memoir *A Mathematician's Apology* proudly proclaims "I have never done anything 'useful'"[8] This avowal prompted a one-sentence review of his book that I've always found amusing, "From such cloistered clowning the world sickens."[9]

Applications and misapplications, attempted proofs and generalizations are not unrelated to the sometimes overweening ambitions of mathematicians.

Chapter 3

AMBITION VERSUS NIHILISM

INFINITY, SETS, AND IMMORTALITY

I certainly would not have brought up the topic of infinite sets were it not a part of my intellectual biography. Specifically, I was obsessed in graduate school with the grandiose ambition of refuting the German mathematician Georg Cantor's famous continuum hypothesis.[1] The hypothesis concerns the cardinality (numerical size) of the set of real numbers, but first let me issue a migraine warning. A touch of abstract set theory lies ahead. You may wish to retreat to a dimly lit room bathed in white noise before proceeding, or you may simply skip the next few paragraphs. As noted, I've included them because of their autobiographical relevance.

The issue is the mind-boggling one of different infinities. Cantor had shown that in a quite precise sense the set of all real numbers (the decimal numbers) is "more infinite" than the set of all fractions (the rational numbers), which is "no more infinite" than the set of all whole numbers (the integers). Employing ingenious so-called diagonal arguments he explored these and other anomalies. Among them is the fact that the set of all subsets of an infinite set is more infinite than the set itself and that therefore there is a whole hierarchy of bigger and bigger infinities.[2]

The question he then posed was both compelling and bothersome: Is there some subset of the real numbers that is more infinite than the set of all integers but less infinite than the set of all real numbers? That is, is there a subset of the real numbers whose cardinality lies between that of the integers and that of all the real numbers? Cantor suspected the answer was no, that the integers, loosely speaking, were the smallest infinity and that the real numbers were the next largest.[3]

Leaving technicalities aside, I would dream of different approaches to the continuum hypothesis that were off the wall and thus, it seemed to me, more likely to succeed than the approaches that had already been explored. As noted, Cantor had conjectured that there was no subset of real numbers intermediate in cardinality between the real numbers and the integers, and his conjecture was subsequently shown to be neither provable nor disprovable from the other axioms of set theory. The task I set myself was to come up with an axiom that was intuitively obvious and from which the continuum hypothesis (or, more likely in my opinion, its negation) provably followed. That is, an axiom that would settle whether there "really" was no intermediate infinity between that of the integers and that of the real numbers or, as I thought, that there "really" was an intermediate infinity. (The scare quotes are meant to suggest the philosophical difficulties inherent in clarifying the status of such weird entities.)

This is not a disquisition on set theory and infinite sets, so I won't go into any more details about my attempts, except to say I looked at various notions of generic sets of reals, of different definitions of random reals, and at a host of ideas involving so-called ultrafilters, structured forcing, and other constructs from model theory. Sometimes I deluded myself into thinking an approach looked promising only to see it fall apart. In short, I never came close to settling on a plausible axiom from which the negation of the continuum hypothesis would follow.

This was a sad disappointment for me. Eventually I settled instead on a much less ambitious project (my thesis was titled "Truth Maximality, Beth's Theorem, and Delta-Closed Logics"), sufficient, alas, to earn me my PhD. My new focus on mathematical logic, probability, and model theory did, however, give me greater warrant to change my approach to the interplay between mathematics and a variety of other disciplines and topics, a project of sorts I've pursued ever since.

Part of the appeal of the continuum hypothesis is that more than most mathematical ideas, the notion of infinity (so-called countable infinity, uncountable infinities, super large cardinal numbers, whatever) has a natural appeal that stems in part, I think, from people's interest in immortality. It also leads to many seeming oddities. For example and as mentioned, Cantor had proved that there were no more rational numbers (all possible fractions) than there were integers. David Hilbert's infinite hotel is a related oddity associated with the integers, the smallest infinity. The hotel has infinitely many rooms, but even if they were all occupied there would still be room for another guest (or guests). Put the new guest into room 1 and move each of the other guests to a room with a number one higher than the one formerly occupied.

In fact there would be room for a countable (i.e., small) infinity of additional guests. Move each guest with room number n into room 2n. That is, the guest in room 5 would be moved into room 10, the one in room 12 to room 24, and so on. This frees up the infinitely many odd-numbered rooms into which the infinitely many new guests can be moved. In a sense this property of the infinite set of integers has been known since Galileo, who pointed out that there are just as many even numbers as there are whole numbers. Likewise there are just as many whole numbers as there are multiples of 17. The following pairing suggests why this is true: 1–17, 2–34, 3–51, 4–68, 5–85, 6–102, and so on.

As the continuum hypothesis suggests, these numerical anomalies involving infinity characterize not just the integers but the real numbers as well. Zeno's famous paradox, which my grandfather once explained to me in a very garbled way, is a classic example. (If he'd explained it clearly, I wonder if I would have been as bewitched by it.) An example less well known, but even more astounding, involves the so-called alternating harmonic series: 1 - 1/2 + 1/3 - 1/4 + 1/5 - 1/6 + 1/7. . . . A natural question suggests itself: what is the sum of this series? We can certainly sneak up on this sum by first finding 1 - 1/2, which is 1/2, then finding 1 - 1/2 + 1/3, which is 5/6, then finding 1 - 1/2 + 1/3 - 1/4, which is 7/12, and so on. The numbers in this sequence of partial sums, when expressed as decimals, are .500, .833 . . . , .583 . . . , ... These numbers get close to .69 and in the limit add up to ln 2, the natural logarithm of 2.

This is already very intriguing, since there doesn't seem to be any obvious connection between the series and logarithms. Much more intriguing, however, is that if we rearrange the numbers in 1 - 1/2 + 1/3 - 1/4 + 1/5 - 1/6 + 1/7 . . . , say, in this way instead - 1 + 1/3 + 1/5 + 1/7 - 1/2 + 1/9 + 1/11 + 1/13 - 1/4 + . . . , then we will get a different sum! Moreover, by an appropriately artful rearrangement of the numbers in the series, we can get can get the sum to be anything we want - 2014, -456,231.57, pi, 42, whatever. Unlike the finite case, the order of the additions and subtractions affects the resulting sum. This is not a bizarre little oddity isolated from scientific practice but is important, for example, in the study of Fourier (math joke: yeah yeah yeah yeah) series and their many applications.

These sorts of considerations bring to mind Laurence Sterne's eighteenth-century novel *The Life and Opinions of Tristram Shandy, Gentleman*, which led British philosopher and mathematician Bertrand Russell to his "paradox of Tristram Shandy."[4] The paradox concerns the raconteur narrator of the book, Tristram Shandy, who, Russell recalled, had taken two years to write the biography of the first two days of his life. Shandy grieved that, at this rate, the latter parts of his life would never be recorded. Russell noted, however, that "if he had lived forever, and not wearied of his task, then, even if his life had continued as eventfully as it began, no part of his biography would have remained unwritten."[5]

The resolution of the paradox also depends on the peculiar properties of infinite numbers. At the rate reported, the account of Shandy's third day would have taken him a year to write as would that of his fourth, fifth, and sixth days. Each year he would have written a full account of another day in his life, and thus even though he would have fallen further and further behind each year, not a day would go unrecorded provided he lived forever.

But life is not mathematics, and one could easily argue that the prospect of immortality is not as appealing, after all, as the notion of infinity. A junior high school English teacher of mine once asked us to write an essay (called a theme in those days) about the greatest torture we could imagine. I decided that living forever would be mine since after having every experience countless times we would long for oblivion. I also mentioned what she characterized as my "disturbing nihilism," that one's task when young was figuring out something amusing or distracting to do in life in order to fill up our time until we die.

I later learned that the gist of my essay has a long history; it is quite similar, for example, to the horror of Friedrich Nietzsche's eternal recurrence, an idea that can be made slightly more rigorous via mathematician Henri Poincaré's recurrence theorem. My English teacher was right. The weariness that infinity elicits in me does bring me to the subject of nihilism.

SELVES AND ABSURDITY

Biographies are about people and the meaning of their lives, alternatively about selves and absurdity, topics I've always found problematic. Even in elementary school I had an outlook that in retrospect can only be described as cheerily nihilistic, and it was coupled with an above-it-all view of the world's follies. My good-boy, equable facade and interest in baseball and other "normal" pursuits hid this from people who simply thought of me as shy.

In my book *A Mathematician Reads the Newspaper*, I mention the meaning or meaninglessness of life in the narrow context of logical meta-levels.[6] (Snow is white versus Mortimer knows snow is white versus Gertrude doesn't even believe that Mortimer knows that snow is white.) As I was there, let me be both schematic and simplistic and stipulate that issues, things in general, really matter in some profound way, or they don't. If they do, that's great. They provide the meaning of life for us. But if they don't, then, as philosopher Thomas Nagel seems to suggest in his famous paper "The Absurd," there's no reason to despair, since even if nothing really matters, it probably doesn't matter that nothing matters. And if nothing matters and that nothing matters doesn't matter either, then why can't we iterate the move at higher meta-levels? It doesn't matter that it doesn't matter that nothing matters and so on and on.[7]

This raises the possibility of an ironic and conceivably happy approach to life. Reasoning formally and again somewhat simplistically, I suspect that the best situation is for people to believe either that things matter at the basic level or, failing that, that they matter on no level. We can have childhood simplicity with its various Santa Clauses for some people or total adult irony and a fearless recognition of meaninglessness for others.

Nagel seems to also make a related point using time rather than meta-levels. (I write "seems" because he both defends and attacks such arguments for life's absurdity.) People sometimes grow despondent when they realize viscerally that what they're doing now won't make a difference in, say, a thousand years. This is probably true, but what will be the case a thousand years from now doesn't make a difference now either. We can iterate the argument once again, but the point remains. That things don't matter doesn't necessarily matter. Colloquially put: relax. It's later than you think. (A Groucho Marx quip is apt: Why should I care about posterity? What's posterity ever done for me?[8])

I don't think such nihilistic attitudes are rare. I noted, for example, a most unexpected expression of such very recently as I walked through a neighborhood not far from my house. From a distance I saw a discarded bed—frame and mattress—leaning upright against the outside of an apartment building. Getting closer I saw "Nothing really mattress" scrawled on the latter.

Of course, pleasant serenity need not lead to indifferent passivity nor to a renunciation of all anger and outrage. At least I hope the only choice isn't between comfort and acceptance of injustice on the one hand and constant distress and struggle with it on the other. A war rages (or maybe just simmers) inside everyone, I suspect, between their inner George Orwell (social justice) and their inner Henry Miller (personal hedonism). One can subscribe to the above in an intellectual way but still be quite conventional about what matters (mattress). I do and I am. Some combination of work (science, business, carpentry, whatever), play, family, friendship, hedonism, travel, food probably constitutes as good an answer as any. Is it enough? The answer, of course, is yesnoyesnomaybenoyesnoyesmaybe.

These musings carry the whiff of a faint Buddhism, which is quite relevant to this book's message on biography. Although this book obviously has small chunks of autobiography in it, I think there is a fundamental falsity that infects all biographies and autobiographies. It is that the very notion of (auto)biography presumes an essential self that is the subject of the story. This self, it should be clear, takes on various traits and attributes and has certain experiences that affect these traits and attributes. As the self travels through time, its present traits affect future experiences and in turn are altered by these experiences. The traits are contingent on the ephemeral experiences, the experiences are contingent on the ephemeral traits, and yet, we assume, an unchanging self persists that is the subject of the (auto)biography. I—my present I, that is—don't think so, but rather, like Buddha and David Hume, I believe such an essential self is a chimera, an illusion. (A case can certainly be made that a more accurate personal pronoun is *we* rather than *I*, and it's a contentious *we* at that. The different parts of each of us holding different beliefs bicker incessantly. I sometimes look back on certain chapters in my life and find that only a deliberate act of careful inference can get me to recognize that I'm the "same" person I was then. I lack an instinctive feeling of really inhabiting that person.)

As suggested above, long before I heard of the two just-mentioned fellows, I remember feeling the same way. Lying on the floor watching *I Love Lucy* or wrestling with my brother, I had the inchoate idea that in an important sense there was no essential difference between me and not-me, that everything was composed of the same stuff and that the air above my forehead and the brain inside it were just patterned differently.

At the time I also noted with childish amusement that *God* was *dog* spelled backward and believed even then that the referent of the former was also an illusion. Spelling out my reasons for this very much later, I wrote *Irreligion* in 2008, a book that cataloged the glaring holes and standard moves and countermoves in all the usual arguments for God.[9] One of the weakest but perhaps most cogent arguments for God is his biography as told in the relevant holy book(s). It's not a big stretch to realize that the self, like God, needs a compelling story. In a sense, selves, like God or gods, are stories. Our lucky breaks, sad disappointments, submissions to duty, attempts at transcendence—in short, our biographies—purport to explain how we got to be the way we are and, as important, how we managed, through our myriad changes, to maintain our seemingly unique and essential selves. Whenever I see the billboard at the Ben Franklin Bridge entrance to Philadelphia advertising Self-Storage (for furniture and the like), my first brief thought is that Self-Storage could be the name of a Buddhist precept urging an at least temporary suppression of our self-expressions and obsessions.

Selves, of course, do exist, but their identity is of a nominal sort, a bit like the identity of a sports team. One can write a history of, say, the Green Bay Packers or the Philadelphia Phillies, but the assumption of an unchanging essence to these teams is harder to take literally and easier to see through. Of course, some complexes of traits are more resistant to change than others, but time wounds all heels and we all change drastically over time. Indeed, we're like Buddhist nuns who go to beauty parlors and get impermanents.

Despite these realizations, I in particular and most people in general are interested in life stories, their own as well as those of others. If I weren't, I wouldn't be writing this book. An appreciation of aspects of Buddhism and the necessarily somewhat narcissistic act of writing a memoir are not the best of friends. A big part of the reason we enjoy biographies, autobiographies, and novels is that stories reinforce our visceral belief in our essential and sovereign selves, and this is how most of us, including me, usually think of ourselves.

But there is another cost associated with taking ourselves too seriously and defining the notion of our nominal self too narrowly. It is that by doing so we can hurt our extended selves in something like the same way we do when we insist on narrow, rigid, and literal interpretations in other domains—when we insist on a "zero tolerance" for drugs, avoid any sort of dirt, or institute relentlessly frequent cancer screenings. Even taking too seriously a disease we have may be counterproductive. (I've suffered from a mild case of Crohn's disease my whole life, and, though an atheist, I tell people that I'm a member of the C(r)ohanim. The reference is to Cohanim, members of the tribe of Levi.) After all, the self/other demarcation is fuzzy at best, as parts of us inhere in bacteria, in nature generally, in other people (family, friends, neighbors, colleagues), in the ambient culture, even, increasingly, in Google and on the Internet. This blurring of the notion of self is, of course, not unrelated to the focus of some Asian philosophy on background and context rather than on foreground and individuality.

The various thought experiments involving teleportation of copies of ourselves, the effect of cutting the corpus callosum in our brains, the closest continuer theory of personal identity, and the like also lend support to this notion of a nominal, vaguely demarcated self. If the average human life span were two hundred years, we moved between planets, uploaded parts of our brains, downloaded parts of the Internet "at will," regularly replaced worn-out organs and tissues with enhanced versions, it would also be more difficult to maintain the fiction of an essential integrity of the self. So think "I."

Dreary or cheery? Let's go with cheery.

THE STORY OF "I"—NEURONS, HALLUCINATIONS, AND GÖDEL

So who is this "I" that we're all so concerned about? It is an entity quite consistent with a nominal quasi-Buddhist notion of self (but not with the absurd beliefs about reincarnation and karma that exists in most versions of Buddhism). It's helpful to think of it as the neuronal self, which is theoretically important even if the details are not (unless one is a neurologist). An abstract account of it that is nontechnical yet I think illuminating comes from computer scientist Douglas Hofstadter's book *I Am a Strange Loop*, which I reviewed in my "Who's Counting" column for ABCNews.com when it came out.[10] I wrote then of a most apt metaphor that helps elucidate how symbolic thought emerges from lower-level neuronal buzz and how, as I put the issue above, the air above my forehead and the brain inside it were just patterned differently. The perhaps obvious joke: yes, we're all airheads.

Hofstadter asks us to imagine a billiard table with countless small interacting magnetic marbles, referred to acronymically as simms. These simms careen around the table, leading to the term *careenium*. Sometimes these simms clump together magnetically and form spherical clusters of simms or simmballs. The simms move around randomly, but these larger simmballs have trajectories that are partially determined by forces external to the careenium, and their movement thus begins to model conditions outside the careenium. Substituting cranium for careenium, individual neurons for simms, and symbols for simmballs, we understand that these symbols and structures gradually reflect a more and more refined representation of the outside world.[11] Note that this awareness does not inhere in individual nerve cells but in the large-scale structures, the symbols, within the brain. Symbols develop for all manner of entities including, at a sufficiently complex level of development, a symbol for an "I." That is, a symbol that represents itself. This symbol is aware of itself and of other objects, people, ideas, desires, fears, motivations, and so on, but not of the neurons of which it's composed. The stories of our "I" symbols are our biographies, the accounts of our turbulent lives, the "I"s of our storms.

This self-referential emergent loop that is the "I" symbol is strange indeed and abstract. Hofstadter argues that neuronal commotion can give rise to high-level symbolic thought. This occurs in something like the way that statements about numbers can be interpreted, via appropriately clever codings and other techniques, as high-level statements about provability, consistency, and the like.[12]

It was Austrian logician Kurt Gödel who demonstrated the latter, that statements about numbers can be thought of as having coded within them higher-level statements that "talk" about themselves and the entire arithmetic system. That is, logical statements in the language of arithmetic can be viewed bifocally as being straightforwardly about numbers, but also about meta-mathematical notions like provability. The distinction, I repeat, is somewhat analogous to that between neurons and abstract thought. One important thing one of these statements says is that the system of which they are a part is incomplete, or if complete, then inconsistent. Since ancient times, mathematicians had assumed that mathematics was complete, that, in particular, every true statement about numbers had a logical proof demonstrating its truth even if they remained unaware of it. It turns out that this is not so.

It's not hard to imagine that a long developmental process has resulted in neuronal movements having coded within them higher-level symbolic patterns that "talk" about the world and themselves. This self-reference and crisscrossing of levels is why we're all, as Hofstadter puts it, "hallucinations hallucinated by hallucinations."[13] It also suggests that human lives and life stories are subject to Gödelian constraints. (Interestingly, Herman Melville in *Moby Dick* anticipated an informal statement of Gödel's first incompleteness theorem. He wrote, "I promise nothing complete; for any human thing supposed to be complete, must for that reason infallibly be faulty."[14])

These high-level mind patterns are not necessarily dependent on the particular physical stuff of the brain. They allow for gradual self-construction and degrees of self-awareness (say, among dogs, chimps, and other animals) and suggest ways we model other people in our minds and thus give them a pale sort of life within us. This self-referential tangle that is an "I," being indefinitely recursive, also allows for an understanding of various language levels, situations, and personas, and their complicated interactions. A long-time interest of mine, humor, in particular, calls on our ability to see alternative interpretations and model others' personalities. But even the verbal and nonverbal cues (raising an eyebrow, changing one's tone of voice, winking) that are present in joke telling are somewhat paradoxical. They say, in effect, "This is unreal," and they are more or less equivalent to the classic "I am lying" or "This statement is false," which is true if and only if it's false.

Between biographies conceived thusly and logical paradoxes there isn't much of a stretch. The Story of I, The Story of U, *The Story of O* (the title of a famous erotic novel by Anne Desclos)—we're all strange and odd, not least in failing to recognize how strange and odd we are.

Although we are in a sense airheads and hallucinations hallucinated by hallucinations, the connections between the symbols in our heads (or simmballs in our careeniums) and the outside world are nevertheless critical. If they weren't, Robert Nozick's famous experience machine would be more attractive than it is. Philosopher Robert Nozick introduced the notion and corresponding thought experiment in his book *Anarchy, State, and Utopia* in which he considered a machine capable of providing us with whatever sort of pleasurable experiences we wanted (or miserable ones, if we were masochists).

Devised by future neuropsychologists (à la the movie *The Matrix*) whose advanced knowledge of our brains permitted them to stimulate appropriate areas and particular neurons, the machines would bring about pleasurable experiences that would be indistinguishable from the "real thing." Nozick asks whether we would prefer these machine-induced experiences to those in real life.[15]

If we were completely hedonistic, we might conceivably choose the machines over real life, but if we were less thoroughgoing hedonists, we might still wish to hook up to them occasionally. After all, isn't watching porn a version of such machines, or even going to the movies or reading novels or biographies? Distraction, vicarious adventures, even self-delusion are, in their place, perfectly defensible choices; choosing to become irreversibly addicted to heroin (a kind of experience machine) is much less so.

In general, one needn't choose between (weak) hedonism and worldly accomplishment, but rather decide, speaking in broad terms, on the relative values of fun and accomplishment. Biographies of thoroughgoing hedonists are, I would guess, relatively rare and distinctly boring long before the thousandth sexual "conquest." Lives of more accomplished people are probably more interesting, albeit, perhaps, sometimes a pain to live.

Chapter 4

LIFE'S SHIFTING SHAPES

PRIMITIVE MATH, LIFE TRAJECTORIES, AND CURVE FITTING

In their book *Where Mathematics Comes From* linguist George Lakoff and psychologist Rafael Nuñez argue that from a minimal set of inborn skills—an ability to distinguish objects, to recognize very small numbers at a glance and, in effect, to add and subtract them—people extend their mathematical powers via an ever-growing collection of metaphors. Some of these shed light on the way we view our life stories.[1]

Watching my grandsons Theo and Charlie play with their toy trains suggests that our common experiences of pushing and pulling objects and moving about in the world lead us to form more complicated ideas and to internalize the associations among them. The size of a collection of Cheerios, for example, is gradually associated with the size of a number. (If my grandsons' scattering of Cheerios is typical, it also leads to the realization that numbers are all over the place.) And combining collections into a single pile is gradually associated with adding numbers, and so on. Another metaphor associates the familiar realm of sticks (or even small Legos) with the more abstract one of measuring and geometry. The length of a stick or a Lego train is associated with the size of a number once some specified segment or single Lego is associated with the number one.

Mathematics may indeed be a more visceral enterprise than people think, as we come to understand most abstract concepts by generalizing, associating, and projecting our physical responses to them.

Scores and scores of such metaphors and analogies underlying other more advanced mathematical disciplines such as probability and statistics can then be developed. Consider the notions of
central tendency—average, median, mode, etcetera. They most certainly grew out of commonplace activities and workaday English
(or other natural languages) words like *usual, customary, typical,
same, middling, most, standard, stereotypical, expected, nondescript,
normal, ordinary, medium, commonplace, so-so,* which in turn grew
out of universal experiences. It is hard to imagine prehistoric
humans, even those lacking the vocabulary above, not possessing
some sort of rudimentary idea of the typical. Any situations or entities or traits—storms, animals, rocks, friendliness—that recurred
again and again would, it seems, lead naturally to the notion of a
typical or average recurrence.

These considerations suggest that both looking back on our
past and thinking of our future lead very naturally to the metaphor
of a trajectory for our lives, and not just any trajectory, but the most
accurate or descriptive trajectory. Thus the proto-mathematics
inherent in my grandsons' toy trains (and my own lovingly cared-
for childhood Lionel trains as well) leads in this natural sense to
the notion of a life's progression and biography. Let me extend this
metaphor a bit further.

A common problem in statistics is finding the curve or surface
that best fits or approximates a relatively small number of data
points in space. Without intending to do harm to our notion of the
richness of human biographies, I note that there is a suggestive
analogy to biography here. It is the similarity between the linking
of data points with the curve or surface that best fits them and the
telling of the story of someone's life by invoking a relatively small
number of remembered events. We infer from the events what must
have linked them. If he did this, that, and the other thing, then he
must have had such-and-such a period in his life or have been such-
and-such a type of person. The events suggest a life's trajectory, but
the narrative constructed from them is usually just an inference

structured by social norms, conventional biases, passing fads, and personal attitudes. Joe Smith is a person who did X1, X2, X3, . . . and Xn, so based on these data points we construct his biography.

We can develop the natural analogy between the mathematical notion of a trajectory and a biography even further. When trying to find the curve or surface of best fit through a set of points in space (physical, psychological, social, cultural, organizational), there are statistical techniques that are used. (There are also ideas about exponential and trigonometric functions and about general Fourier series that are germane if one wants to get more technical.)

We don't want to have too many points far away from the curve or surface, just as in telling a life story one doesn't want too many events that are incongruous with the basic arc of the story, such as the great humanitarian as a teenager spitting in customers' food or Benjamin Franklin's essay on flatulence, titled "Fart Proudly."[2] And just as the statistical techniques used tend to erase the individual data points in favor of the summarizing equation of the best curve or surface, so our biographies often tend to sacrifice the out-of-character events, incidents, and minor intrigues in favor of the basic outline. That's probably true to an extent in the autobiographical segments included herein.

The length of a trajectory or life is usually considered a crucial fact about it, which is probably why we first check the age of people when reading obituaries and why it's usually given in the headline or first sentence. It's always "Waldo Jenkins, 89," or "Amy Winehouse, 27," and not "Waldo Jenkins, 5'7", 176 pounds."

Incidentally, an idea from probability helps clarify a common misunderstanding about life spans. If the average life span is, for example, 80, and someone is 72, it's not the case that that person can expect only 8 more years of life. His conditional life span, given that he's lived 72 years, might well be 93. Likewise, if you were to roll a pair of dice, the probability of getting a sum of 12—two 6s— is 1 in 36, just 1 of the 36 possible outcomes for the dice. But the conditional probability of getting a sum of 12, given that you know you got at least an 11, rises to 1 in 3, the three possible outcomes on the two dice being 6,5; 5,6; and 6,6.

For most people obituaries are a particularly conventional sort of biography, reducing even the most complex individuals to a stylized litany of birth, early schooling, college, career choice, marital history, noteworthy achievements, and survivors. Almost as conventional are most autobiographies of sports and political figures (in contrast to those of autobiographies of scientists and writers, which tend to take place largely in their heads). Whoever the subject might be, the addition of a couple of vivid incidents would, if nothing else, make the biography more interesting. There are too few that produce the reaction "It's truly amazing that somebody who produced the wonderful X could also have done the execrable Y (or vice versa)." Nobel Prize–winning author V. S. Naipaul comes to mind because of his reported serial physical abuse of women.

Viewing life stories in this metaphorical way also sheds some light on a common phenomenon: people often drastically reevaluate their lives after a few emotionally significant events. How can we model the general trajectory of our life, which is to say the curve or surface that best fits its salient events, being significantly changed by only a few such events? One way is by introducing or just emphasizing another dimension with respect to which our life's trajectory looks quite different. That is, assume that for a while the importance of some dimension of our life—wealth, for example—increases. But then we introduce some other dimension—a relationship or a child, for example—and we note that as more time passes, the importance of the wealth dimension decreases and the importance of the other dimension increases. Some such reevaluations are quite common as one's life horizons recede.

If the discovery of the new dimension or criterion with which to evaluate our lives occurs late in life and if we give it great weight, then our evaluation of our life will shift dramatically as will the curve or surface that best fits it. This can be stated in terms of the rates of change with respect to time of these various dimensions or criteria, but the idea is, I hope, clear as a metaphor. "Conversions" of one sort or another provide examples, a couple of which I'll discuss later.

Conceiving of our lives as the curves or surfaces of best fit passing through an appropriate space provides us with another bit of mathstuff that describes and, to an extent, constitutes us. Our trajectory captures our narrative, the narrative of ourself. Moreover, it gives a metaphorical meaning to a number of phrases such as "over the hill," which feeds into our visceral understanding of the parabolic shape of trajectories followed by rocks and balls and, as suggested, by us. In fact, if we don't push the analogy too hard, basic notions from calculus, somewhat simplistically applied, can help us demarcate the rough divisions of most lives. Youth can be grossly defined as that part of our lives where the first and second derivatives of our development (indicating the rate of change and the rate of change of the rate of change, respectively) are positive, corresponding to increasingly rapid growth. Middle age is more problematic. It is that part of our lives where the second derivative of development is negative while the first derivative remains slightly positive or slightly negative, corresponding to slow growth or slow decline, respectively. Old age begins either when the first derivative first becomes quite negative, corresponding to rapid decline or, as I'd prefer, when the first derivative remains negative, but the second derivative becomes positive again, corresponding to a very slow decline. Would that we approach death only asymptotically and never arrive there.

The notion of a curve or surface of best fit also suggests that short lives in which the latter part is truncated make better stories because it's easier to impose a coherent narrative upon them. Someone who has moved around a lot, has been buffeted by the vagaries of the world, has had several disparate careers and obsessions, and has interacted with sequentially distinct sets of friends and colleagues will have an interesting life, but not an especially coherent story line (or an easily comprehensible arc). Yet another reason to find the curve or surface of best fit in statistical applications is prediction and retrodiction of future and past events. This is less important in biography, but even here we often want to predict what a still-living biographical protagonist will do in the future or retrodict what he or she might have done in the past. Pitfalls such as statistical overfitting (models that are too complex and may only be describing random noise rather than anything significant) have biographical analogues as well.

I should note that this book, focused as it is largely on me and mathematics, is admittedly somewhat solipsistic. (I originally intended to title it *The Book of John 3.14*.) Not from lack of regard, but I've ignored almost entirely the rich tapestry and interplay between my trajectory and the trajectories of my friends, family, and colleagues, whether from Milwaukee, Philadelphia, New York, or many elsewheres. (Again the reason is not that they are unimportant to me but merely the narrow focus of this endeavor.) My father-in-law, whose kindest appellation for me was "jerko," would merit a few chapters all by himself. I'm not a novelist, and capturing with appreciation this intricate woven texture over time is beyond me. Trying to give a minimalist account of just myself is already too difficult.

Having mentioned the word *novelist* and liberally invoked several mathematical metaphors, I would like to cite Leo Tolstoy as an illustrious codefendant in this particular use of mathematical metaphor. The above metaphors and others to come later may seem a bit much, but, though rare, such figurative speculations are not unheard of. Tolstoy famously proposed in *War and Peace* that calculus be used to model historical change (biographical change writ large). He wrote, "A modern branch of mathematics having achieved the art of dealing with the infinitely small can now yield solutions in other more complex problems of motion which used to appear insoluble."[3] He further argued, "Only by taking infinitesimally small units for observation (the differential of history, that is, the individual tendencies of men) and attaining to the art of integrating them (that is, finding the sum of these infinitesimals) can we hope to arrive at the laws of history."[4] Even if "modern branch" is understood to encompass the calculus of variations, field theory, and other more recent developments, I'm much less sanguine than the appropriately titled Count Tolstoy about there being any such laws, but I applaud his invocation of mathematics in such a context.

Less of a stretch than Tolstoy's: Playing around with Lego trains leads naturally to the notion of life trajectories.

THE ENVIRONMENT AS A PINBALL MACHINE, THE QUINCUNX OF LIFE

The trajectories of our lives through cultural/psychological spaces also suggest the topics of chaos theory and nonlinear dynamics. These now somewhat familiar notions arise when we examine certain systems—economic, ecological, physical, even personal— whose important variables are linked by nonlinear equations, not straight-line relationships. In the case of economic systems, for example, interest rates have an impact on unemployment rates, which in turn affect revenues. Budget deficits, trade deficits, and the total national debt also influence interest rates as well as consumer confidence and the stock market. These rates and totals as well as economic problems in other countries affect almost all economic indices, each of which feeds back on the others, reinforcing (or weakening) them and being, in turn, affected by them. This convoluted complexity suggests that the answer to many, if not most, economic questions is "Duh, I don't know," an answer no politician would ever dare give. Even human-made computer code worked on by many programmers over many years is so complex as to be often unpredictable.

Similar connections characterize interactions in nature, a more poetic rendition of which is due to the Buddhist monk and poet Thich Nhat Hanh, who writes, "If you are a poet, you will see clearly that there is a cloud floating in this sheet of paper. Without a cloud there will be no water; without water, the trees cannot grow; and without trees you cannot make paper. So the cloud is in here. The existence of this page is dependent on the existence of a cloud. Paper and cloud are so close. Let us think of other things, like sunshine. Sunshine is very important because . . ."[5] What he doesn't add is the unpredictable nature of this ecological complexity.

Complicated nonlinear interactions such as these characterize any modern economy or any natural ecology and often give rise to the butterfly effect. This is the phenomenon whereby a tiny event, say, a butterfly flapping its wings somewhere, can lead to an unpredictable cascade of events resulting over time in a major event, say, a tsunami in some distant locale. The term derives from the work of meteorologist Edward Lorenz modeling the dynamics of weather; the geometrical shape of his model resembled a butterfly.[6]

Or think of a pinball machine with an irregularly warped base and numerous randomly placed flappers in which tiny differences in the initial speed, angle, and spin with which two balls start their trajectories soon results in their following quite different paths. Balls bouncing off the pegs and obstacles in directions only minusculely different, one will soon hit a peg or bump the other misses or vice versa, after which the balls' trajectories diverge radically.

The contingency of people's paths and actions suggests the question of how effective any of us, even BBBs (brilliant billionaire Buddhas), might be in solving a given societal problem. The question is, of course, too vague to answer, but maybe no BBB or no small group of them could be sufficiently smart, rich, and compassionate to effect such change. Or maybe—a depressing thought— the law of unintended consequences would prove more powerful than any bunch of BBBs, and their impact would be generally seen as negative. Or, more likely perhaps, they wouldn't do vastly better, or even better at all, than the same-size bunch of randomly selected ordinary people. After all, no one can herd butterflies. These considerations also undermine a similar story I liked as a kid, that of a completely selfless, thoroughly beneficent, nondescript, almost invisible person who nevertheless was responsible for an inordinate amount of good in the world—not Maxwell's thermodynamic demon but a kind of Maxwell's angel, opening the door to the good and just, closing it to evil. This, like Santa Claus, is an appealing story, but, alas, it also is a fantasy.

The delicate dependence of the pinballs' trajectories on very minor differences is similar to the dependence of one's genetics on which a sperm cell haphazardly zigzagging along makes it to the egg first. Although harder to formally model as part of a nonlinear system, the disproportionate effect of trivial events also comes to mind—the accidentally deleted e-mails, missed flights, serendipitous meetings, and odd mistakes that shape and reshape our lives. Just as there is ample evidence that some economic, ecological, and physical systems are subject to the butterfly effect, so, I think, is there reason to believe that we ourselves as well as our relationships are, to a considerable extent, nonlinear systems sensitively dependent on initial conditions. Our biographies are thus, to an extent, an attempt to string together all our contingent traits and interactions into a coherent, yet quite contingent, narrative.

For this reason among many others we are all instantiations of mathematical notions and theorems; in this case the math-stuff is nonlinear dynamics, but it could be almost any branch of the subject. We're something like the prime-numbered thirteen- and seventeen-year-old cicadas, the golden-spiraled sunflowers and pineapples, and the tessellations of honeycombs, but vastly more complicated. We have instantiated within us the properties of numbers, the distributions of probability theory, the rules of logic, the beauty of calculus, the whole pantheon of mathematical patterns and ideas. We have the geometry of the genome's DNA, the network theory of our brain's connections, and the sinusoidal rhythms of our circadian bodily functions. Mathstuff permeates us and our lives, and, if you're someone who subscribes to a Platonic conception of mathematics, you might even be tempted to call it the "divinity" in all of us, a mathematical idealization that allows us to think of ourselves as gods and goddesses rather than as dying animals.

Let me move from Plato's realm to a couple of less ethereal examples of sensitive dependence and contingency—the first a humdrum biographical illustration of nonlinear dynamics, the second a matter of considerable public importance.

My two brothers Paul and Jim, sister Marilynn, and I grew up in the same home, albeit no doubt in different micro-environments. We're similar in many ways (not all as trivial as our liking *The Three Stooges*), but we also have very different interests. More generally, there are a number of factors relevant to any set of siblings developing differently. Early labeling such as the wild one or the studious one or the warm one and the repelling effect on siblings of the so-called narcissism of small differences are two such factors, but bouncing off different pegs in the pinball machine that is life shouldn't be underestimated and, as noted, is yet another reason to be very dubious of biographers' (or even autobiographers') general pronouncements.

We tend to think we've arrived at our present station largely by dint of determination and hard work, but as my father used to say, we're all just farts in a windstorm. Less graphically put, we're all parts of various systems—familial, professional, societal—and these systems impact on us and direct our paths as if we were pinballs whirling through the quincunx of life. Nevertheless, we should heed the aforementioned title of Benjamin Franklin's essay, "Fart Proudly." That is, we should embrace our contingency even when it's unpleasant.

The second example of an extremely significant, decidedly unintended result of a relatively tiny event is nightmarish, at least for me. It concerns the role I played in getting George W. Bush elected president in 2000. That I was the butterfly whose fluttering cascaded into Bush's election still pains me. I had written an op-ed for the *New York Times* titled "We're Measuring Bacteria with a Yardstick" in which I argued that the vote in Florida had been so close that the gross apparatus of the state's electoral system was incapable of discerning the difference between the candidates' vote totals. Given the problems with the hanging chads, the misleading ballots (in retrospect, aptly termed butterfly ballots), the missing and military ballots, a variety of other serious flaws, and the six million votes cast, there really was no objective reality of the matter.[7]

Later when the Florida Supreme Court weighed in, its chief justice Charles T. Wells cited me in his dissent from the majority decision of the rest of his court to allow for a manual recount of the undervote in Florida. Summarizing the legal maneuverings, I simply note that in part because of Wells's dissent the ongoing recount was discontinued, the case went to the US Supreme Court, and George Bush was (s)elected president.[8]

Specifically, Judge Wells wrote, "I agree with a quote by John Allen Paulos, a professor of mathematics at Temple University, when he wrote that, 'the margin of error in this election is far greater than the margin of victory, no matter who wins. Further judicial process will not change this self-evident fact and will only result in confusion and disorder.'"[9] (Incidentally, the CNN senior political analyst at the time, Jeff Greenfield, cited the quote in his book on the 2000 presidential election, *Oh, Waiter! One Order of Crow!*, and wrote "the single wisest word about Florida was delivered not by a pundit but by mathematician John Allen Paulos." I doubt, however, that Greenfield thought it was reason to stop the recount.)[10]

I was surprised and flattered, I admit, that I was cited by the judge but also very distressed that my words were used to support a position with which I disagreed. Vituperative e-mails I received didn't help. Many were angry that I would support Bush. Some were clearly demented. With all due respect to these correspondents and the esteemed judge, I believed and still believe that the statistical tie in the Florida election supported a conclusion opposite to the one Wells drew. The tie seemed to lend greater weight to the fact that Al Gore received almost half a million more popular votes nationally than did Bush. If anything, the dead heat in Florida could be seen as giving Gore's national plurality the status of a moral tiebreaker. At the very least the decision of the rest of the court to allow for a manual recount should have been honored since Florida's vote was pivotal in the Electoral College. Even flipping a commemorative Gore-Bush coin in the capitol in Tallahassee would have been justified since the vote totals were essentially indistinguishable.

Historical counterfactuals are always dubious undertakings, but I doubt very much that the United States would have gone to war in Iraq had Gore been president. I also think strong environmental legislation would have been pursued and implemented under him. Was I responsible for Bush's presidency? No, of course not; butterflies can't be held responsible for the unpredictable tsunamis that in retrospect can be traced to their fluttering and to a myriad of other intermediate events. Still, every once in a while, the guiltifying thought that the unwarranted Iraq War was my fault does occur to me.

BIOGRAPHIES AND THE TEXAS SHARPSHOOTER

One unavoidable problem associated with the idea of a life's trajectory is illustrated by the old story about a Texas rifleman who was thought to be an expert sharpshooter. The shooter's secret was that he would fire shots at a barn or a sign and then when many of the bullet holes happened to cluster together or form some sort of pattern, he would draw a target or other shape around them and then strut about claiming to be a sharpshooter.

Given that any life has countless incidents, situations, and characters, it's clear that biographers might be especially prone to the Texas sharpshooter fallacy. (The fallacy reminds me of the observation that meteors always seem to land in craters.) More generally, in this era of Big Data, the National Security Agency (NSA), and state and corporate surveillance, we must always be wary of such cherry-picking of data. What makes the fallacy especially common in biographies is that once a reputation takes hold, almost the only events that are recorded are those consistent with it. Bad guys seem to do only bad things; compassionate people perform only considerate acts; mass killers are always said to be loners; parents are usually characterized as family men or devoted mothers; George Washington and Abraham Lincoln were always completely honest; and so on. It's as if the Texas sharpshooter's bullets make no holes when they deviate too far from the target or conventional trajectory.

A possible recent example came to light on the hundredth anniversary of the birth of British mathematician and computer scientist Alan Turing. Turing did seminal work on the eponymous universal Turing machines and theoretical computer science in general and performed invaluable lifesaving service during World War II on cryptography. His achievements can hardly be overstated. Literally. Despite them, he was convicted of indecency because of a homosexual affair in the '50s and opted for hormone treatments rather than prison. He died of cyanide poisoning a couple of years later. After his death he became something of a symbol of gay oppression, and biographers wrote of his being harassed, despondent, and suicidal. Philosophy professor Jack Copeland has suggested, however, that there was no evidence for the standard story of depression and suicide and that the cyanide poisoning was accidental.[11] We'll never know the truth, but there is certainly a tendency to prefer narratives, such as the first, that comport with our expectations.

More generally, biographers can vehemently insist on a biased view of a person that is unsupported by much evidence. The person is assumed to be manipulative, saint-like, laid-back, greedy, or whatever. But such views don't exist in isolation and exert a distorting effect on other matters.

An analogy from geometry is helpful. Were a physicist so inclined, he could assert without inconsistency, but contrary to Einstein and others, that space was Euclidean and flat rather than non-Euclidean and curved. If he did so, however, he would have to account for astronomical phenomena that can be explained quite simply and naturally in a non-Euclidean framework. The benighted physicist would be compelled to introduce fictitious forces and accelerations to save his assumption that space was Euclidean and flat. Which geometry/physics combination to use is to some extent a matter of convention, but some conventions are better than others. In the same way, to insist that a person is, for example, always manipulative when by most accounts this is not the case requires that one weirdly interpret the actions of some people, question the motivations of others, or assert the naïveté of still others. Insisting, for example, that a politician, say, President Barack Obama, is an evil socialist, as some Tea Party members do, requires that they undertake wholesale distortions of their standards of judgment.

The so-called conjunction fallacy, or Linda problem, suggests a related pitfall of just-so stories with little evidentiary value. The more details in a biographical anecdote or even in an everyday conversation, the more plausible and engaging the account becomes. Alas, it also grows less probable. The reason is simple: the more details there are, the less likely it is that the conjunction of all (or most) of them is true.

If, for example, Senator Jones seems to be an extremely devoted and happily married family man who lives modestly in a small house, which is more likely: (a) Jones accepted an illegal campaign contribution from a supporter or (b) Jones accepted an illegal campaign contribution from a supporter and used it to pay for his daughter's expensive medical treatments? Despite the more coherent story the second alternative begins to flesh out, the first alternative is more likely. For any statements, A, B, and C, the probability of A is always greater than the probability of A, B, and C together since whenever A, B, and C all occur, A occurs, but not vice versa.

As with the Texas sharpshooter foible, approaches to biography or even everyday storytelling that depend on the conjunction fallacy are quite common. It's interesting watching how some people effortlessly embroider, exaggerate, gerrymander, and invent details to concoct a compelling little anecdote out of the sparsest and most ordinary of incidents. Munchausen syndrome, whereby healthcare providers and/or patients exaggerate reports and add false details to obtain sympathy, attract attention, or portray themselves as heroes, is an extreme example.

My predilection has usually been just the opposite. I find excessive enthusiasm suspect and often feel compelled to report neutral facts that undermine the tendentious slant of any story I read and thereby drain it of much of its drama. I can be an irritating killjoy. This deflating habit is one reason why I would make a very bad novelist or biographer. My wife, who has also taught at Temple University for many years with more vivacity than I can muster, is quite different. Before her teaching career, she wrote romance novels and occasionally exploited my debunking psychology in doing so. She would ask what I would do or say in some situation, and she then made sure her male protagonist did or said something radically different. Given any conflict, I generally look for possible misunderstandings that might contribute to it and then search for common ground, not an approach conducive to much bodacity.

This is not always the best strategy, as the unlikely story of the nonfunctioning guillotine and the condemned engineer shows. Other potential guillotine victims were released when the blade didn't fall. The engineer, however, once his head was placed on the block, noted, "Oh I see the problem. The rope has a kink and has come off the pulley on the left side and can't lift the blade." The problem was duly corrected and the engineer beheaded.

Clearly misunderstandings should sometimes not be dispelled, but my professorial bias leads me to think that on the whole dispelling them is good cognitive hygiene. More specifically, biographies would be better with more tentative analyses and fewer apocryphal events. Biographies would also be better off, albeit occasionally less entertaining, if they were less extreme, whether in a hagiographic or a demonizing way.

Like accounts of fictional characters, biographies, too, often depict people who are in so many ways more extraordinary than the people we meet in everyday life. Biographical subjects seem to react more, emote more, and have more well-defined opinions, motives, and goals than the people we know. They're focused, and their opinions, motives, and goals lead more frequently to decisions and actions. Because of this, they're often more predictable than people in real life, more likely to struggle with others for what they want, less likely to dither and vacillate. Most everything about biographical subjects is more clear-cut and logical. Wishy-washy, conflicted, and uncertain they are not. They also appear to be much more self-governed and less buffeted by the vagaries of chance.

People are, of course, quite different from each other in countless subtle and not-so-subtle ways, but it's hard for me, a confirmed ditherer and vacillator, to credit these stark unidirectional differences between the proverbial people on the street and the subjects of (auto)biographies. Often only a few lucky breaks are sufficient to make an average shmo biography-worthy, and then his or her otherwise complex but superficially nondescript story will be retroactively recast in a more heroic mold. (Incidentally, this use of "average shmo" is more consistent with the median than it is with the mean or mathematical average of a group of people.)

Chapter 5

MOVING TOWARD THE UNEXPECTED MIDDLE

A FEW TOUCHSTONE MEMORIES

Autobiographies differ in a number of ways from biographies, one of which is the inclusion in the former of certain primal memories not credibly accessible to a biographer. These memories usually involve times when people consider themselves to have been more pure, less troubled, more fully engaged. Being in the moment may be easier when one is young, either very young or an adolescent.

I once wrote a very bad short story, happily unpublished, with a related theme: the tension between the protagonist's childhood memories of wholeness, belongingness, aliveness, and his present distracted, alienated, fragmented state. The protagonist attempted to somehow "use" the memories (i) to moor himself, (ii) to exorcise and transcend them, and (iii) to enable him to combine cynicism, disaffection, and jadedness with the feelings engendered by these touchstone memories and become a whole adult. He failed.

Whether this is an absurd romanticizing of childhood or not, the topic brings to mind this very early remembrance of mine. I was three or four years old and sitting in a corner of my maternal grandmother's apartment in Chicago. No one was in the apartment except my grandmother, who was quietly cleaning the bedroom. It was very still as I slowly and intently built a little house with my Lincoln Logs. I loved the dark-brown logs, the green roof shingles, and red chimney. The sunlight was streaming through the window and illuminating the dust motes in the air, and as I contemplated my little house and the dust, I ate one dried fig after another from the ring of figs my grandmother had given me. I was happy to be completely alone and exquisitely conscious of myself and of the silence, the floating dust, the sweet taste of the figs, and my log cabin construction. I felt as if I understood everything, and even now, looking back on it, I think in some sense I did. Everything was of a piece and in peace. Heraclitus's quote comes to mind: "Lifetime is a child at play, moving pieces in a game. Kingship belongs to the child."[1]

Other related memories provide a little fuller context. I am the first-born grandson of Greek immigrants and was born on the Fourth of July. Because of the former and perhaps a bit of the latter, I was shamelessly cosseted by my grandparents, being proof in a way that they had made it in America. My first few years—I'm trying, probably unsuccessfully, to steer clear of the ever-lurking haze of nostalgia—seemed magical, from the late-afternoon sun glinting off the red bricks and rusted black fire escapes of the nearby apartment buildings to the pattern of whorls on my wooden closet door as I fell asleep, from eating toast with the crusts cut off with my then very young parents to the strangely appealing smell of the sewer in the alley behind our building.

Again the romantic coloration may be sneaking in. As one ages, one tends to remember remembering and, to iterate, to remember remembering remembering, and so on, and, like the children's games of telephone and whisper down the lane, small mis-rememberings creep in and the connection between the present remembrance and the original experience grows more and more tenuous as does the connection of one's present self to one's earlier selves. With that proviso, I can even remember the taste of the tomato and basil salads I ate with my paternal grandfather at night on the rooftop of the apartment building in Chicago where we lived and of being mesmerized by the way the light from the streetlamps flickered up through the trees.

My memory of the apartment stretches from the rooftop to the porch below. Every summer night as the sweltering weather eased a bit, a motley assortment of friends, relatives, and neighbors would gather on the sidewalk and porch steps and talk and joke and occasionally argue while I sat unobtrusively somewhere nearby alternately listening, daydreaming, and playing with some toy. I remember thinking that most of what was being said was a little silly, but somehow this made me feel oddly happy, very secure, and a little superior. My mother was beautiful (still is for a ninety-year-old), and my father was a baseball player (who, as mentioned, did play professionally for a bit). Things were very, very good. Why can't I get back there?

It turns out that occasionally I can at least glimpse this state of being safe and wholly in the moment in other people. Such memories were activated one early morning in, of all places, Bangkok not too long ago. The stimulus was a young girl of about eighteen on a motorbike. She had three young kids sitting behind her on the bike and was driving down a crowded, narrow *soi* (side street) until she saw some fragrant flowers on a tree branch overhanging the street. She stopped for a full two or three minutes to smell them. Her foot was extended to keep the motorbike from tipping over, and occasionally an impatient driver would swerve out to pass her and come horribly close to running over her foot, but she continued to smell the flowers and pulled down the branch farther so that each of her children on the bike could smell them too. I could see her talking and laughing with them. Finally, she picked up her foot and rode off. I noted that she and her children were helmetless and their hair was blowing in the breeze.

Her obliviousness, and this is the point, was truly astounding, simultaneously frightening and charming. She was totally there and exemplified the trade-off between the payoffs of prudence and the pleasures of immersion. Being in the moment, one does not appreciate risk or understand coincidence, which are notions that are usually foreign to those who view the world as a whole and not as a combinatorial mix of objects and events that occasionally produce an oddity of one sort or another.

Possessing a PhD in mathematical logic, I'm a little pained to realize that thinking logically often makes being in the moment more difficult. To think logically, after all, is not to let irrelevant details affect your conception of the "structure" of a situation; in other words, to look to what's the same or similar among a large class of situations. But this is also to lose what is distinctive or different about these situations. Too much schematics often results in bland, colorless maps and flowcharts instead of rich, living, vivid, childlike perceptions. Logical thinking, though absolutely indispensable, can lead to a hardening of the categories and a loss of vivid perception. Differences are washed out. (A partial antidote I've found: don't use the words *every*, *some*, *if . . . then*, etcetera. But do use concrete words, active voice, and present tense. Avoid too many adjectives, adverbs, and dependent clauses.)

This conflict between enjoyment and accomplishment, between a mindful/mindless hedonism and a busy industry is quite general. I'm writing this in Singapore after just returning from Thailand, and the relative balance of these two desiderata seems quite different in the two countries. (Singapore is a rich, quite orderly country and its people especially hardworking, especially my students at Nanyang Technological University.) One other piece of advice that is often given when this conflict is mentioned is that we should live each day as if it were our last. This is a rather fatuous suggestion that, if followed literally, would result in millions of people quitting their jobs immediately. Losing their incomes would in many cases lead to living their last day sooner than they otherwise would, but that is probably not what the counselors envision. Oddly, the opposite advice to live each day as if we had eighty years ahead of us is also rather appealing.

There is something very right about both suggestions.

LUNCH, GOOD NIGHT, AND MY PARENTS—
MILWAUKEE IN THE 1950s

The mathematical structure of such idiosyncratic remembrances is perhaps somewhat akin to a mist of fine dust. It lacks the purposeful directionality of more public aspects of a biography. Nevertheless, indulge me as I continue in this vein for a bit in order to lay down at least a minimal memoir, my personal dusty mist.

Moving from a warm ethnic neighborhood in Chicago when I was five to what I felt was a sterile outpost in suburban Milwaukee ninety miles to the north was wrenching. Many memories of my early years in Milwaukee were clouded by a feeling of not belonging, of being small and insignificant. We lived outside the school district, so, unlike my classmates I had to take a public bus to elementary school, which fact contributed to my alienation. My brothers and I used to play touch football on our postage-stamp-sized front lawn, which somehow seems pathetic in retrospect. I longed for the demographically dense tenement where we lived in Chicago with my doting grandmother.

My young mother missed Chicago and my grandmother as well and was no doubt quite overwhelmed, her hands full with my two younger brothers and sister. Nevertheless, I felt neglected. I remember going to school in an ill-fitting plaid shirt, torn pants, and with my lunch in a large brown grocery bag that I tried with limited success to squash into something closer in size to a conventional lunch bag. Inside the bag was my lunch of a jelly sandwich (the jelly, invariably grape, leaking through the Wonder Bread that was so soft it took finger prints, and visible through the plastic of the bread we'd just finished), a couple of broken cookies, and a bruised or semi-rotten apple. This was hardly traumatic, but neither was it exactly conducive to being accepted, much less popular. For whatever psychological or other reason memories of the '50s seem like a movie shot through fog—muted colors and almost no sound. Denver, where we went for the summer to stay with my grandmother, who had moved there from Chicago, was different; it had vivid colors and was full of noisy talk. For the most part, however, the '50s were a time of dreamy unconsciousness and a feeling of smallness.

Regarding the latter, I think that, whatever the period, most autobiographies, memoirs, and even little sketches of early childhood like those above paint the protagonist as small and sensitive. This obvious geometric fact naturally goes unmentioned, but few children think of themselves as being large and invulnerable since, simply put, they never are. Because we all start out small it takes us a good while to realize that we have little reason to feel especially aggrieved. It takes us even longer (for some of us, never) to realize that we can ourselves sometimes be aggrievers. Our tendency to always be the aggrieved party, the blameless moderate, is cleanly stated by comedian George Carlin: "Have you ever noticed that anybody driving slower than you is an idiot, and anyone going faster than you is a maniac?"[2] Because of a hard-won feeling of egalitarian symmetry, I thoroughly dislike autobiographies that paint the protagonist as being regularly victimized. The sentiment is shared by many people ranging from Carlin to Franz Kafka, who wrote, "In the struggle between yourself and the world you must side with the world."[3]

As a very trivial example of a tendency to be controlling and my not-so-occasional status as an aggriever, I recall the totalitarian ritual I instituted whereby my younger brothers Paul and Jim and I said good night to each other. We slept in a loft-like, large room and engaged nightly in momentous discussions of how I'd hidden the scrambled eggs I detested in the cuff of my pants, why the rag we wiped our faces with hung in the garbage, what were the new tortures we had devised for Hitler, and perennial jokes involving the detonating of our "warheads," which I'll leave undefined. After a time, however, I, the oldest, would autocratically decide that a sufficient amount of pre-sleep talking had taken place and loudly proclaim, "No more talking time." Then I'd say "Good night, Paul and

Jim," after which Paul had been told to say "Good night, John and Jim." Finally Jim knew to say, "Good night, John and Paul." If any of us said anything to interrupt the flow or even laughed, I insisted that we had to do it all over again. Sometimes Paul added a funny comment, which, after a rebuke from me, necessitated starting from the beginning, and occasionally Jim would fall asleep so quickly that Paul and I would have to wake him up in order to properly go through the good-night procedure. I felt the perfection of three "good nights" was ruined by the perversion of a fourth remark or, worse, only two.

A telling little vignette about my parents or, more specifically, their differences is not inappropriate here. As I wrote in *Once Upon a Number*, I stayed home sick one day when I was in elementary school.[4] As soon as everyone had gone for the day my mother turned on the hi-fi (sounds so retro, like a Ford Model T) and swirled through the house as she watched *The Helen Morgan Story* and listened to *Madame Butterfly*, and torch songs of unrequited love. I liked watching her as she danced and talked on the phone and did her housework in a sort of romantic haze unconnected, it was clear, to my father. Then my father came home, his suit creased, his tie loosened, a characteristic lopsided grin on his face. After rumpling my hair affectionately and lighting an ever-present cigarette, he said, "Hi John boy. Spahn's pitching tonight. The Braves are going to come out of their slump. You'll see." Much later, after only thirty-six years of marriage, they divorced. It shouldn't have been surprising. My father, to my mother's chagrin, was not cool. My mother, to my father's, was not warm.

LOGIC, JOKES, AND ADULT LIFE AS AN UNEXPECTED PUNCH LINE

Over time, of course, my parents changed, as did their relationship. The relationship was better in the beginning else it wouldn't have even had anything beyond a beginning. More generally, people, the situations they find themselves in, and descriptions of these situations evolve in countless ways and give rise to vastly different and unpredictable stories. In fiction as well as biography almost identical plot lines—boy meets girl, faces obstacle one, obstacle two, whatever, boy gets girl—can develop all sorts of variants and complications and will yield quite different novels. Hence, of course, the word *novel*.

In something like the same way, if I may abruptly shift tone, the punch lines of good jokes are not predictable from what their setups might suggest. Sometimes if the setups are long and extended, quite different punch lines will make sense. Often these punch lines depend on a nonstandard interpretation of the joke's setup. Even answers to simple riddles can number in the hundreds. What's black and white and re(a)d all over? An embarrassed zebra, Santa Claus coming down a dirty chimney, a wounded nun, a skunk with diaper rash, and so on and on.

Some mathematical logic lies behind the above observation about jokes and the previous one about life stories. I discussed the former in my first book, titled *Mathematics and Humor*, where I considered various theories of humor and sketched the formal structure of many types of jokes and riddles, including the following sort.[5] Joke-teller: In what model (that is, under what interpretation) are axioms 1, 2, and 3 true? Listener: In model M. Joke-teller: No, in model N. Ha ha. The incongruity of the different interpretations for the axioms accounts for whatever humor the jokes possess.

During the course of writing the humor book I exchanged a number of letters with the prolific mathematical writer Martin Gardner, who knew a surprising number of dirty jokes of this type. Question: What goes in stiff and dry and comes out soft and wet? Answer: Chewing gum. A G-rated example of mine from the book: Consider the story of the young man who signed up for an online dating service. He stipulated that he wanted a potential spouse who enjoyed water sports, was gregarious, was comfortable in formal wear, and was distinctly shorter than average. The service recommended a penguin. Pedantically spelling this out, I noted that the young man's requirements play the role of axioms and that the natural interpretation or model of these axioms is a young woman who satisfies the conditions mentioned. The penguin and its lifestyle provide the axioms with a different and unexpected model. Incidentally non-Euclidean geometry can be thought of as a kind of joke, a nonstandard model of Euclid's axioms (minus the parallel postulate).

In this sense (and others), a person's life may also be thought of as a kind of joke. Its punch line is that his or her later behavior and persona is at odds with what the person's earlier self might have predicted. Schematic biographical joke: A person who had such-and-such early experiences as a child would become what sort of person as an adult? Answer: This sort of person. Biographer: No, not that sort of person, but this surprisingly different person. I sometimes think of this when I look back at myself as a kid and fail to see a strong connection between the person I recall and my present self, which makes my life, like yours, sort of a joke.

I remember, for example, as a seven- or eight-year-old (before I got smart in the fifth grade) arguing with my father about my desire to go directly into the major leagues and skip college. He gently suggested I could play baseball after college and let me hold on to the illusion that I'd be good enough to even make the high school team. A few years afterward I experienced the aforementioned highlights of my "baseball career": a game-winning home run over a neighbor's backyard fence and a diving, pants-ripping catch in center field. As a kid I also loved very bad World War II movies full of references to "japs" and "krauts," set up mock battles with my toy soldiers and model tanks, and hung carefully constructed fighter jets over my bed.

Now, among countless other differences, I hate war movies and action movies in general and have lost my fervent interest in home runs, diving catches, and sports in general. Also I have a visceral distaste for people of any ethnicity, religion, social identity, or demographic group that parochially, incessantly, and proudly talk about being members of their tribe. As some (many? most? all?) people have, I've made sometimes conscious efforts to move beyond and in some cases reject outright the attitudes prevalent in my familial and social background. Moreover, I will not eat a lunch packed in a large supermarket grocery bag ever again.

Incongruous, strangely funny, nonstandard models of the adult continuation of our early lives are part of what's at the root of the sense of astonishment we sometimes feel when we look back. I'm reminded of the compelling first line of Anne Tyler's novel *Back When We Were Grownups* (indirectly referred to in the introduction): "Once Upon a Time, there was a woman who discovered she had turned into the wrong person."[6] Why did I, so aware now of everyday risks, feel so secure sitting under my mother's rickety ironing board on Saturday mornings watching cartoons. Why was I so excited when I received a Valentine in the mail from Miss Gerlach, my third-grade teacher, and then why was I so sad when on Monday morning I discovered everybody in the class had. How did I become, to invoke a common stereotype, a Volvo-driving, latte-drinking cynic? Such changes, like so much else, suggest that personal identity is a nominal, evanescent thing and not an unchanging essence that adheres to us throughout our lives.

The realization of how large the difference between our childhood selves and adult selves is somewhat muted by hindsight bias, a tendency to assume that we always knew what we know now. In particular, we tend to continually refashion our life stories to better correspond to our present selves, so we're less likely to see how much we've changed. This phenomenon is not unrelated to a common response to the announcement of new research findings in medicine or the social sciences. People often scoff, "Oh, that's obvious. I knew that." The problem is that their response would often be the same if the research yielded the opposite findings.

A related point regarding biographies: any collection of statements at all, not just those describing early life, will admit of many different and often incongruous interpretations or models. The statements constitute the bare bones of biographical fact that can then be dressed up in a multitude of ways. Consider how political parties and campaigns take the facts (or at least some version of them) and spin them in almost unrecognizably different ways. The same, of course, is true of potential biographies of a person. Even given the same indisputable facts, it's not especially hard to come up with debunking versions of biographical subjects regarded as heroes or glorifying versions of those regarded as villains. The biographies of Steve Jobs are a good recent example.

Of course, some facts are more constraining than others, but in general the facts under-determine the story. The problem is a bit like trying to find a cubic polynomial through only 2 points in the plane; lots of very different curves will do. As the Löwenheim–Skolem theorem in mathematical logic shows, the undisputed (first-order) axioms of arithmetic or set theory have nonstandard models, not just the intended ones of the familiar whole numbers or the intuitively understood interpretations of basic set theory. The fact is that most sets of axioms, say, those for some abstract algebraic or topological structure, will admit many surprising models, punch lines to the axioms as it were, just as our later lives are punch lines to our earlier selves.

Bottom line once again: Whatever splendor or darkness we experience in our lives, our life stories can, in a sense, be considered jokes.

MEMORIES AND BENFORD'S LAW

In writing some of the reminiscences for this book I noted a strange lacuna in them. I was able to retrieve fewer memories from the middle part of my life. I could do so, of course, but only with more effort. The reminiscences were disproportionately from my childhood and early adulthood or from relatively recent times. It turns out that this is not unusual.

Studies have shown (a useful locution, but too often a weasel clause) that the number, vividness, and perspective—whether pre-reflective and immediate or reflective and removed—of autobiographical memories that people can easily recall change over time. From adolescence and early adulthood there is a large increase in the number of our easily retrievable memories, what is called the "reminiscence bump," brought about, it's thought, because changes in identity, the making of salient decisions, and the establishing of an adult persona are most common then. Relevant is the observation that older people, say, aging baby boomers named John, know the groups popular when they were in high school or college but are clueless about today's popular music.[7] (Presumably people who undergo life-changing transformations and refashionings when older—say, a sex change in their sixties like Bruce/Caitlyn Jenner—will experience another reminiscence bump then.) Moreover, memories from long ago are more likely to be reported with the removed perspective of an observer than more recent ones, which are generally reported with an immediate pre-reflective perspective and detail. The emotional content of the memory also plays a role, as does a person's degree of self-awareness and a host of other psychological factors including, as I'll discuss later, psychologist Daniel Kahneman's distinction between the experiencing self and the remembering self.[8]

These considerations seem to have played a role in the kinds of memories related here in this book. Of course, a self-selected sample of one (me) generally is quite convincing to that one (me).

Interestingly, as I intimated above, the number of memories people can easily recall from their midlife is often relatively low. Overall the rough story seems to be: vivid stories from early life, the reminiscence bump, midlife unconcern with autobiographical memory, and then many memories both immediate and reflective as one gets older.

I'm not a psychologist, and the area of study seems to contain many open questions and even more vague definitions. Nevertheless, these variations in autobiographical memory bring to mind Benford's law. This relatively recent mathematical theorem states that in a wide variety of circumstances, numbers as diverse as the area of river deltas, physical properties of chemicals, figures in a newspaper or magazine, populations of small towns, and the growth of money in an account all begin disproportionately with the digit "1," whatever units are used. Specifically, these numbers begin with "1" about 30 percent of the time, with "2" about 18 percent of the time, with "3" about 12.5 percent of the time, and with larger digits progressively less often. Less than 5 percent of the numbers in these circumstances begin with the digit "9."[9]

The general theorem is hard to prove, but an easy, special case of this phenomenon involves the growth of money in an account. If you put $1,000 into a bank at, say, 7 percent, the amount will remain below $2,000 for a number of years (leading digit 1), remain below $3,000 for fewer years (leading digit 2), remain below $4,000 for even fewer years (leading digit 3), and remain below $10,000 (leading digit 9) for the fewest number of years before repeating the cycle of initial digits. (Note that this is in stark contrast to many other situations—say, where a computer picks a number between 1 and 9,999 at random—where each of the digits from "1" to "9" has an equal chance of appearing as the first digit.)

Without going into details, I note that Benford's law has been used to detect fraud in accounting reports and IRS returns. If there are too many leading 5s, for example, and not enough leading 1s in these forms, Benford's law is violated and deceit becomes a likely possibility.

My general skepticism of biographies brings me to the relevance of an as yet nonexistent Benford-like psychological law to autobiographies. If too many memories from a certain period of one's life or too many with an observer's perspective or too many vivid remembrances or too many of the wrong kind for a given period in one's life are related in an (auto)biography, then deceit or at least exaggeration becomes a likely possibility, just as too many leading 5s and not enough leading 1s suggests fraud on a tax return. Of course, numbers are much more clear-cut than the various types of memory recall, but if the science of the latter progresses, it along with many other techniques could be routinely used to flag suspicious auto-biographical reminiscences. We could also use Benford's law itself if many numbers are included in the (auto)biography. Other mathematical results such as so-called power laws (the basis of Pareto's 80–20 principle whereby 80 percent of the effects flow from 20 percent of the causes) and adaptations of the algorithms used to detect plagiarism could also be invoked if we're inclined to question the veracity of biographies. The NSA's metadata might help as well. (A few hundred calls weekly between Waldo and Myrtle might belie their denial of an affair.)

So might movies. An expert on memory, the psychologist Daniel Schacter wrote, "In the 1980 presidential campaign, Ronald Reagan repeatedly told a heartbreaking story of a World War II bomber pilot who ordered his crew to bail out after his plane had been seriously damaged by an enemy hit." He continued, "The press soon realized that this story was an almost exact duplicate of a scene in the 1944 film *A Wing and a Prayer*. Reagan had apparently retained the facts but forgotten their source."[10]

Of course, we can and should also rely on common sense. We should be especially dubious of sharp and fresh details in the reminiscences of old people. They're sometimes made up since the real ones have vanished from their memories. Often their sources, as with Reagan, are also misattributed. Furthermore, people who could challenge their accounts of greatness may be gone in one sense or other. "Who's around to say that I didn't score seven touchdowns at the playground that day," is a claim my brother Paul once made, much to the continued amused dubiousness of my brother Jim and me, but to no one else.

Ah, the nuanced interplay between narrative and number.

Chapter 6

PIVOTS—PAST TO PRESENT

KOVALEVSKY, PREDICTION, AND
MY GRANDMOTHER'S PETTY LARCENY

Before I indulge a bit more in reminiscence, I'd like to digress and offer up the story of Sofia Kovalevsky for reasons that will become clear.

Sofia Kovalevsky (Kovalevskaya) was a brilliant Russian mathematician of the nineteenth century who died at forty-one. Her story, like that of mathematicians Évariste Galois, Srinivasa Ramanujan, and Frank Ramsey, all seminal mathematical thinkers who died too early, is a sad but romantic one. Youthful accomplishment and promise cut short usually is. Despite Kovalevsky's cossetted early life followed by the daunting antifemale cultural environment she faced later, she managed to make major contributions to partial differential equations. The field is the multivariable generalization of ordinary differential equations, which relates the rates of change of some quantity with respect to other quantities. This may sound fairly banal, but the subject of differential equations has clarified some of the chief glories of modern civilization, among them Newton's laws of motion, Laplace's heat and wave equations, Maxwell's electromagnetic theory, the Navier–Stokes equations for fluid dynamics, and Volterra's prey-predator systems. (The latter is related to the analysis of the love affair in *Gone with the Wind* mentioned earlier. The repeated and alternating rise and fall of the prey and predator species—one up, the other down—can be shown to be somewhat analogous to the repeated and alternating love and hate attitudes of Rhett Butler and Scarlett O'Hara.)

Kovalevsky's dissertation, "The Theory of Partial Differential Equations," helped expand inquiry into differential equations to include nonlinear dynamic phenomena, many independent variables, and what would come to be called chaos theory.[1] She was a colleague of such other groundbreaking mathematicians as Karl Weierstrass and Henri Poincaré, who in addition to his work on chaos, topology (Poincaré conjecture), and qualitative differential equations, established before Einstein the mathematical framework for special relativity theory. This late nineteenth-century focus on dynamic phenomena and many complex interacting variables is perhaps a mathematical reflection of the many cultural, economic, and political changes then reshaping the world. Cultural and societal issues do sometimes affect the types of problems mathematicians work on. Note the relatively sudden rise of interest in the mathematics of networks, which is not unrelated to Twitter, Facebook, and other common networks.

The reason I write about Kovalevsky, however, relates to a brief biography of her by Nobel Prize–winning author Alice Munro in her collection of stories titled *Too Much Happiness*. In her masterful story Munro tries to integrate the mathematical and the personal aspects of Kovalevsky's life. The story itself struck a chord with me because, very roughly speaking, Munro's story attempts to do for Kovalevsky what in a small way I've attempted to do here for myself.[2]

Kovalevsky's mathematical, personal, and social concerns were interconnected. One example is her work on quantities (functions) dependent on several variables and their dynamic behavior as well as her unhappy love affair and its continuation, which was shaky and contingent on many factors. She herself wrote a memoir, *Recollections of Childhood*, and an intriguing autobiographical novel, *Nihilist Girl*, whose titles alone resonated with me. Therein she commented: "In what manner should we act in the future to make our common life happier? Mathematically we could have stated this question in this manner: given a definite function (in this case our happiness), which depends upon many variables (namely our monetary resources, the possibility of living in a pleasant place and society and so forth)—in what manner can the variables be defined so that the given happiness function will reach a maximum? Needless to say we are unable to solve the problem mathematically."[3]

I've made no seminal contributions to mathematics, nor have I won a Nobel Prize in literature, nor have I led an inordinately dramatic life. Nevertheless as noted I have also tried to yoke the disparate realms of the personal and the mathematical in an unusual way that I hope will prove interesting and perhaps useful to readers.

To that end, let me leave Kovalevsky, whose childhood bedroom, incidentally, was decorated with mathematical symbols, and resume the story of my own, very different nonfiction nihilist boy. Important components of my or anyone's early life are models, not in the mathematical sense but in the conventional one of parents, grandparents, and their contemporaries. Given rapid social and technological change, these are people from whom you're increasingly likely to differ, and who thus provide a measure of how much you've changed. This musing is prompted by the general considerations above (adult behavior incongruous with childhood experiences) and, in particular, by memories of my relationship with my diminutive Greek grandmother, whose bedroom, incidentally, was decorated with relatives who all appeared to me to be aliens and several hundred years old. She was kind and gentle and I loved her dearly, but just as she would have been surprised, were she alive, at how I "turned out," I find it slightly shocking to recall some of her traits. One of these was a tendency toward petty larceny.

Because of this trait, going somewhere with my grandmother was often an adventure. She and my grandfather emigrated from Greece soon after their marriage, and after raising their family and owning various restaurants in Chicago (where I consumed gargantuan portions of rice pudding), they retired to Denver, Colorado. For years I would visit them—or just her after he died—during summer vacations. They bought four small apartments in a one-story building on Denver's east side, and my grandmother would go there with obsessive frequency to water the lawn, sometimes mumbling something about the grassy bliss around the apartments reminding her of Greece. The claim seemed far-fetched to me even then.

She would take the bus from her house on Jasmine Street to the apartments on Xenia Street and insist to the driver that I was under twelve and therefore was permitted to ride for free even though I was about fifteen and near my adult height of 5'11". The force of her personality and threatening glare were enough to intimidate both the bus driver and me. He and I shrugged and sometimes exchanged complicitous smiles, and I tried to hurry down the aisle to get away from the frowns of the people near the front who had heard the embarrassing interchange.

When we got to the apartments she would drag out the hoses and sprinklers and leave me to read the *Rocky Mountain News* or *Denver Post*, George Gamow's *One, Two, Three ... Infinity*, or, if it was very hot, to amble down to the motel at the corner of Colfax and Xenia to get grape sodas from the soda dispenser there. The labyrinthine metal path that the soda had to pass through to be retrieved from the machine always made the anticipation of its coldness a little keener. Picking up a soda that's fallen into a trap door on modern dispensing machines is much less enchanting.

The lawn watering accomplished, we'd walk ten blocks to a dilapidated house overgrown with weeds and nearly covered with a large weeping willow tree where my grandmother had found a mysterious old woman even shorter than her 4'10". The woman sold raw honey, *meli* in Greek. My grandmother seemed oblivious to the stench of the house and claimed that this was real honey, not the sort they sold at King Super, the local supermarket. She illustrated her distaste for the contemporary version with a characteristic squinching of her nose and a quick upward movement of her head, which was also her normal way of saying no.

Then, if I was lucky, we took the bus back to her house on Jasmine. (Why is the specificity of place names so evocative and redolent even without the prompting of a fragrant name like

Jasmine?) If I was not so fortunate, the embarrassment of the bus ride, to which I was accustomed if not inured, was supplemented by a visit to the King Super. It may not have carried the kind of honey she liked, but it did have a lot of other products she wanted and that she thought were exorbitantly expensive. The result was not hard for me to predict, and I usually insisted on waiting outside the supermarket, where she would meet me with her loot. Once she took a large ham out from under her shawl-covered armpit and displayed it with pride.

Even a trip to the local A&W drive-in root beer stand was tinged with my grandmother's lawlessness. There would be six of us in the car, six orders of root beer, but only five mugs when the poor attendant came back to pick up the tray. This only happened a couple of times because she was shamed and badly outnumbered by the five law-abiding citizens in the car.

Looking back on these incidents, I can see elements of the present me in the past me, but lacking a comparable foresight, I doubt if, at the time, I would have been able to fathom much of my present self. What can I trace forward? I tend to be just as absent-minded as I was when I would walk home from kindergarten during recess. And my basic honesty I may owe to my father's easygoing decency, as well as, oddly, to my grandmother's petty thefts.

Of course, projecting traits into the future is problematic and seldom falsifiable. Plausible alternative explanations will always abound because the underlying "psychological science" drastically under-determines the traits in question, as philosophers from Pierre Duhem to Willard Van Orman Quine have observed.[4] The same is true of "culinary science." The success or failure of a complicated recipe can be attributed to almost any ingredient(s) in it as well as to variations in the amounts, in the times, in the methods of cooking, in the order of ingredient addition, and so on.

Whatever effect, psychological or culinary, my grandmother may or may not have had on me, whatever her lapses in honesty, I loved her. Still, I didn't always approve of her, and, were she in the habit of making pronouncements, I doubt that I would agree with many of them. But she generally confined herself to cooking for me and telling me what a wonderful kid I was, actions and assertions with which it was difficult to disagree even for an adolescent yearning to be independent.

And I still like honey, soft drinks, and reading newspapers (and their various online descendants) and dislike anything to do with lawns and, in fact, lawns themselves.

TURNING POINTS, ACADIA TO KENYA

Biographies generally focus on "turning points" in a person's life. V was a W-ish fellow until X happened, after which he became Y-ish and obsessed with Z. Lives certainly have turning points like X, but is there ever a turning point in one's life that is largely internally generated, a decision that is not a direct response to outside events but largely due to internal ones alone? The adverb *largely* is a bit of a weasel word since it can mean just about anything, and all turning points are at least minimally affected by both internal deliberations and the external environment. Nevertheless, are there many such points where the outside effect is tiny, all but invisible? Do people, after long consideration, whether conscious or not, ever decide to make a sharp break to go that way rather than this? This is an impossibly vague question bordering the tricky philosophical terrain surrounding free will and determinism, but my suspicion is that an appropriately qualified "sometimes, but seldom" is the rough informal answer.[5]

As I've suggested, our belief that we are captains of our ships and largely autonomous beings is quaint, but likely only true if the ship is a raft bobbing about in a raging, stormy sea. This isn't to say that our psychological and physical makeups are irrelevant. They're not. Decisions are filtered through us, but the environment, being vastly more complex than we are, is, in a sense, the primary decider.

But let me sidestep the theoretical issue of free will and its definition and simply note that if we decide something, it has to be, informally speaking, a real choice that might have gone the other way. This, I suppose, is the interpretation of the well-known joke about Jean-Paul Sartre working on his book *Being and Nothingness* in his favorite Paris café. He tells the waitress that he'd like more coffee, but with no cream. The waitress replies, "Monsieur Sartre, we are out of cream. How about more coffee, but with no milk? We have milk."[6]

This isn't necessarily a dumb waitress joke. It may be a brilliant waitress joke. One of the central ideas in Sartre's philosophy is the distinction between real choice and the mere appearance of such. Arguably Sartre can't genuinely choose to have more coffee with no cream, because cream isn't available, but he can genuinely choose to have more coffee with no milk, because milk is available.

That all said, I've had many of the conventional turning points in my life—points on a life's arc where there are sharp twists. Marriage and fatherhood in particular have been life-giving, life-altering, and life-enriching. (Some sentiments can be both saccharine and true. These are certainly not meant to devalue other choices or lifestyles.) I won't elaborate, but the words *dad* and *husband* tend to change one irrevocably. So in a far lesser way did stopping for a two-day visit to Bar Harbor and Acadia National Park on our way to Quebec. The fortuitous detour began my family's long, annually renewed love affair with Maine's beautiful Mount Desert Island and the hiking, biking, and climbing it affords us. Even these, however, were quite context-dependent and did not grow ex nihilo but developed naturally out of certain situations (age, location, culture, etcetera).

My experience in the Peace Corps was even more a consequence of external events, namely, my belief in the imminence of being drafted to fight in Vietnam in 1969. Teaching in Kenya seemed a much superior option. I was in graduate school in mathematics at the time and, despite the excitement and turmoil of the '60s, had lived a rather conventional life. Teaching and traveling around Kenya, Tanzania, and Uganda made many topical issues much more vivid and visceral than they'd been before. These included political and economic matters (an intense exposure to real spirit-sapping poverty), personal ones (risky sex, in several senses, in particular with the wife of a local chieftain with whom I regularly played cards), social mores (horror at the memory of the headmaster of my school using the lower half of a hollowed-out elephant leg as an umbrella stand), and scientific issues (wildlife, conservation, the Great Rift Valley). I even met President Jomo Kenyatta, often considered the founding father of Kenya, who, glancing at my '60s-era shoulder-length hair, asked me if they had barbers in the United States.

As I related in *Innumeracy*, another turning point of sorts was my reading some of British philosopher, mathematician, and social critic Bertrand Russell's writings in high school. I was already very interested in mathematics, logic, and scientific induction, the "scandal of philosophy," and became intrigued by the idea that one could do mathematics and still be engaged in a variety of other intellectual and social endeavors. After reading Russell's *History of Western Philosophy*, I wrote him something of a fan letter when I was a freshman at the University of Wisconsin. In it I asked him what precisely was the logical mistake that Hegel had made. To my great surprise he answered my letter and stated that Hegel's "dialectical argument against relations is wholly unsound. I think such a statement as 'A is west of B' can be exactly true. You will find that Bradley's arguments on this subject presuppose that every proposition must be of the subject-predicate form. I think this is the fundamental error of monism. With best wishes, Yours Sincerely, Bertrand Russell."[7]

More exciting than Russell's answer was the fact that this illustrious ninety-two-year-old philosopher who at the time resided in Wales would take the time to respond to a fan letter from a college kid in Wisconsin. My letter to him perhaps reached him at a low moment. Whatever the reason I was thrilled. A few years later in graduate school I wandered through the university bookstore and noticed the third volume of Russell's autobiography had just arrived and that a copy of it was splayed open to pages 252–253 on a display stand. I picked it up, and my name caught my eye. Russell had included his letter to me amid letters to a pantheon of twentieth-century personages ranging from Nehru and Khrushchev to T. S. Eliot, D. H. Lawrence, and Ludwig Wittgenstein, inducing in me a strange sense of proximity to these historical figures.

Publication of my book *Innumeracy* was also a turning point, which I'll get to later, but all this turning is dizzying, so let me remark on a related general phenomenon: sudden and wholesale changes in personality, goals, and outlook brought about by different material circumstances, religious conversion, sexual infatuation, and whatever. In the example of religion such changes often seem to be a sharp movement toward a more fundamentalist version of the person's formerly nominal religion. I know indirectly of people like this and find their conversions somewhat unfathomable and, as recent news stories demonstrate, frightening. In Thailand, where I've lectured several times, I've also seen firsthand a secular phenomenon that is more understandable, but oddly similar: sexual infatuation and the subsequent personality changes it brings about. It's unsettling to witness (or hear about from my friend Christopher G. Moore, a Canadian novelist living in Bangkok) middle-aged (or sometimes much older) pillar-of-the-Peoria-community types become so quickly almost unrecognizable. Quiet, earnest family men, they suddenly act giddy as they fulsomely pile on the adjectives to gush over their thin, little, pretty Thai girlfriends with their flashing white smiles and long, lustrous black hair—young women who, the usually temporary converts aver, are adorable, sweet, spirited, and playful. And, it should be added, generally quite poor. These sudden changes are, it seems to me, a usually futile attempt to choose a different, more vibrant self, more consistent perhaps with their youthful fantasies.

I mentioned above points in a life at which there is sharp twist. Such religious and sexual transformations are more than twists or bends but rather are points where, in math talk, the life arc is no longer differentiable but is a spiky fold. Change and turning points are what biographers and people in general are most interested in. There is not much to say if life proceeds smoothly and uneventfully along.

I'll end with a common set of usually faux turning points: milestone multiple-of-ten birthdays, thirty, forty, fifty, and so on. To underline their artificiality and lessen the dread that often accompanies them, I sometimes point out to people that their age can be expressed less traumatically in a numeral system with a different base. Happy 40th Birthday, for example, becomes Happy 34th Birthday in a base-12 system—3 twelves and 4 ones. For greater reductions in age-angst, we can express Happy 50th Birthday as Happy 32nd Birthday in a base-16 system (hexadecimal)—3 sixteens and 2 ones.

The same holds with seemingly significant dates such as the turn of the century. The year 2000 in a base-2 system is 11111010000, and 2015 turns out in base 2 to be a palindromic year equaling 11111011111, while 2048 turns out to be the epoch-changing 100000000000.

PAST ACCOMPLISHMENTS VERSUS PRESENT POTENTIAL

How much weight do you give to past accomplishments and how much to present abilities when assessing a person's life? Of course, the weights differ in different contexts, but near the extremes are the illustrious person with a prolific and stellar background but who suffers from dementia and the extraordinarily bright college student with magnificent potential but who still toils toward an initial substantial achievement. How does a biographer or an interested onlooker balance these two extremes—the contemporary and the historical—when writing about someone living?

The question occurs to me because I've met a number of very eminent people, some of whom were, at the time I met them, at least, superficially unimpressive. A bit disconcerting, these meetings should nevertheless have been encouraging rather than disillusioning. "Wow. Isn't it wonderful that someone so unprepossessing should have achieved what he or she has." (One notable exception was Isaac Asimov, who on the spur of the moment composed a funny and suggestive limerick about my wife whom he had just met. He had also recently read my book *Mathematics and Humor* and riffed extemporaneously and wittily on a couple of its jokes and commented insightfully on my use of the mathematical theory of catastrophes to model certain joke types.)

What counts as a significant accomplishment will vary over the lifetime of a person as well. For example, I invented a variant of the Rubik's Cube that I called About Face, which was suggested by a book I had once seen that consisted of drawings of faces that became other faces when inverted. Moreover, twisting the cube replaced the parts of one face with parts of another face. I thought it was an appealing idea and patented it. Alas, no toy company was interested since the Rubik's Cube fascination was fading. At the time I thought my About Face was no mien accomplishment. Now, not so much.

For a more general example, consider someone who has mastered most of the algorithms in college-level mathematics. He will, if he lives long enough, see his mastery grow increasingly unimpressive as simple programs and apps will be able to accomplish the same calculations more quickly and reliably. At least in this way technology does seem to lead to the devaluing of old people. In the other direction, however, the last practitioners of a dying, formerly quite common art, say, being a sushi chef, will see their accomplishments valued much more highly than when they were young.

Whatever one's reaction to these issues, the distinction between past accomplishments and influence and present abilities and potential is a natural one. It brings to mind the notion of "physical entropy," which was originally employed by the physicist Wojciech Zurek and others to clarify the problem of Maxwell's demon and related issues in classical thermodynamics. (Science writer George Johnson's *Fire in the Mind* lays out the notion of physical entropy more accessibly.)

Zurek defined physical entropy to be the sum of the complexity, appropriately quantified, of (what's already been revealed about) an entity and the surprise, appropriately quantified, inherent in (the yet-to-be-revealed aspects of) an entity. Imagine a long but finite sequence of 0s and 1s. As more and more of it is revealed, the complexity of the revealed part grows, while the surprise at the yet-to-be-revealed part shrinks. This technical notion can also be used metaphorically to model the way the weights accorded to past accomplishments and future potential change over time. The entity relevant here is a person's life; complexity of his or her past grows with time, while surprise at his or her future shrinks. (The two notions of complexity above are due to Greg Chaitin and Claude Shannon, respectively.) As a life progresses, the algorithmic complexity of the recorded past (Chaitin's notion, which I will discuss later) grows, while the surprise and potential of the future (Shannon's notion, which is a measure of uncertainty) shrinks.

One aspect of this metaphor that I like is that the notions involved are in the same conceptual ballpark as the second law of thermodynamics, which British chemist and novelist C. P. Snow famously used to illustrate the gulf between scientific elites and literary ones, the latter presumed not to understand the significance of the second law. For this reason I find its historical resonance and its marginal relevance to biography especially satisfying. It's marginal perhaps, but still worth more than yet another account, for example, of Sylvia Plath's life, poetry, and suicide. Incidentally, Plath's interest in astrology suggests she probably didn't understand any thermodynamics.

Finally, let me note that the biographical use of such precise physical analogues is problematic, but even more so is evaluating a potential biographical subject's accomplishments. This is not a mechanical exercise. As with choosing the friendliest colleges or the most lovable cities, we can by an appropriate choice of criteria, measurement protocols, and weightings make someone more or less worthy of a biography. Nothing is wrong with this as long as we are aware of the inevitable element of subjectivity present in every such choice.

A different sense of thermodynamics is the topic of the next chapter.

Chapter 7

ROMANCE AMONG TRANS-HUMANS AND US CIS-HUMANS

ROBOROMANCE AND THE END OF BIOGRAPHY

C. P. Snow brings to mind other issues spanning the so-called two cultures. One is, What becomes of biography when humans are enhanced with robot-governing programs and apps, becoming in effect human-computer amalgams and thus more literally mathstuff? Does it become largely a matter of describing software and hardware upgrades, version 4.2b and the like? Will we all be nothing more than universal Turing machines (after the famous computer scientist and mathematician Alan Turing) with different capacities, fleshware, and speeds? What if some of the readership of the biography operates with a different version of the software than that of the protagonist? What happens when memories can be tweaked, changed, implanted, or even deleted by drugs or noninvasive surgery? If we wait until next Wednesday or Thursday, this may become a reality. Even now standard psychological techniques can be used to induce false memory syndrome. How will real-time access to huge data sets and the never-ending cascade of technological developments change our ideas of what it is to be human? And, given the importance of a private sphere to human and perhaps even pseudo-human thriving, what are the consequences of surveillance that can't be avoided or rigid formalisms that can't be overridden?

These are profound issues that undermine our traditional notions of humanity and hence of biography, but let me focus instead on a lighter but related note. What might we say about relationships or dating in which two such enhanced individuals interact? That is, what might a roboromance look like? An example of one such story that comes to mind is simultaneously retro, futuristic, and a little silly but is germane in its presuppositions. It concerns a kind of antiseptic, logical seduction. Suppose a software-enhanced human male were to brag about his various system upgrades and his familiarity with many technical matters while flirting with a human female whose software and hardware enhancements compelled absolute and scrupulous honesty. (The sexes in the scenario can, of course, be altered as well as permuted, and the scenario itself can reflect other sorts of interchanges, say, ones involving a financial transaction. The stereotypical psychologies and assumptions can also be changed.) After a while, he asks her, "Will you solemnly promise to give me right now your telephone number if I make a true statement and, conversely, not give me your number if I make a false statement?" Feeling that this is a flattering and benign request and not so strange given the man's background, the woman promises to give him her telephone number if and only if he makes a true statement.

The man then makes his statement: "You will neither give me your telephone number now nor will you sleep with me tonight." Flustered a bit, she thinks through her options. She realizes that his statement is either true or false and that, if it is true, the statement says that she won't give him her telephone number. But if his statement is false, she must not give him her number because of her promise to give it to him only if he makes a true statement. Thus, whether his statement is true or false, she won't give him her number. That is, she won't give him her number under any circumstances.

But if she also refuses to sleep with him, his statement becomes true, and this would require her to give him her number, which she can't do. The only way she can keep her promise is to sleep with him so that his statement becomes false. The woman's seemingly innocuous promise ensnares her. We can all be similarly ensnared by bad software. (A briefer alternative is the following question: If I ask you for sex, will you give me the same answer as the answer to this question?)

Fortunately, I suspect that the class of unenhanced people for whom these seduction approaches would prove effective is still rather tiny. The paradox may be a bit goofy, but, as suggested, it does have a point. When human-computer amalgams become more common and difficult to disentangle, "seductions" of this sort might conceivably be part of the resulting creatures' biographies. My guess is that post-singularity (i.e., after the attainment of significantly trans-human capabilities through technology) sex as well as life in general will not be particularly romantic and will be almost certainly unfamiliar. What, for example, would be the consequences of an orgasm button on some descendant of the smartphone or countless other developments we can't come close to anticipating? (Supply your own prescient puns, puzzles, and paradoxes here.)

To reiterate, it seems a soupçon of privacy and perhaps of ignorance, hypocrisy, and duplicity as well is essential to most of life's activities and some of its best moments. After all, less than noble motives can sometimes lead to wonderful adventures. But if we were to abjure entirely these human deviations from logic and probability and think of ourselves as decision and inference machines, what would our biographies consist of? Merely accounts of our responding to new events by updating our increasingly Brobdingnagian databases, both personal and social? Such machines would depend on Bayes' theorem, a well-known statistical technique mentioned earlier, which tells us how we should modify our estimates of the probabilities of events in the light of new information. We all differ in the way we attach probabilities to events, and we differ even more in the probabilities we assign to the associations between events. This tangled network of probability estimates is, in a sense, a map of our minds and interacts with the new experiences we have and with the old stories that we're constantly editing. Using Bayesian techniques, however, our individual beliefs gradually get washed out and begin to conform to those of others' beliefs. Revision of our subjective probabilities generally brings our personal viewpoint, no matter how unreasonably idiosyncratic at first, into better agreement with torrents of new, more objective evidence.

Appealing (or appalling) as these and future developments in genetics, nanotechnology, neuroscience, and other disciplines may be, they won't make for enthralling biographies of remarkable individuals, in part because individuals won't be seen as that remarkable. In fact, if vastly greater than human intelligence ever comes about, biographies will cease to make much sense because the notion of humans, at least as we know it, will have ceased to make much sense. Who will be the last biographical subject? The question maybe should be, Who will be the last autobiographical subject? Whoever it is, I imagine him or her taking a swig of something strong and announcing to no one, "Bot's up!"

CHOOSING A SPOUSE, MEETING MY WIFE, SHEILA

Logical seduction aside, a number of other ideas from game theory, probability, and various mathematical disciplines tangentially are relevant to the issue of choosing a partner. Imagine, for example, you're in the market for a spouse (market and spouse—a decidedly unpleasant juxtaposition of terms), and you can reasonably expect to meet N potential suitors—candidates for marriage—during your life. There will be different values of N for different people. A well-known exercise in probability asks us to assume that the candidates can be numerically ranked from worst to best (according to your tastes) and that at any point, as you date them sequentially, you can stop your search, forego meeting the rest of the total estimated N candidates you're likely to meet, and marry the one you're dating. Or, of course, you can reject him or her and continue on your quest, giving up the possibility of reconsidering that person at a later time.

Given these assumptions, it can be shown that you will maximize your chances of choosing your heartthrob, the candidate who is best for you, if you reject the first 37 percent of the N candidates you're likely to meet and then marry the next one who is better for you than all previous candidates. Using this strategy your chances of ending up with your heartthrob is also about 37 percent. (Incidentally, 37 percent happens to be approximately 1/e, where e, once again, is the base of the natural logarithm.)[1]

This approach is not nearly as poetic but is a bit more mathematical than Elizabeth Barrett Browning's "How do I love thee? Let me count the ways." Although quite simplistic in its assumptions, the "Let me count the suitors" approach does offer some wise counsel: don't opt for the first few possibilities, but don't wait too long either.

Happily, this is not the way my wife, Sheila, and I met. We were both graduate students at the University of Wisconsin and had seen each other a couple of times on campus. She was from New York, I from Milwaukee, and we were superficially quite different. Thinking I looked like the son of her building superintendent whom she disliked, she hurried away when she saw me. A bit later we were both attending an anti–Vietnam War rally near the campus that had turned into something more ominous. The police arrived and began tear-gassing people in what seemed an indiscriminate manner. The resulting chaos found Sheila and me next to each other with a tear gas canister at our feet. We both ran toward a building where we headed for the water fountain. We got there simultaneously, and I held the faucet open for her to rinse her eyes. After I did so as well, we laughed with relief and began talking, not about the rally but about ourselves. She had just returned from working on an Israeli kibbutz and, as mentioned, I had just returned from Kenya with the Peace Corps where I had gone to avoid the draft. Needless to add, perhaps, we got along famously.

A couple of days later we continued our conversations at an Italian restaurant, where another minor explosion of sorts furthered our budding relationship. The waiter was talking with a friend for ten or fifteen minutes, and we were sure our order was sitting on the kitchen windowsill getting cold. A Coke bottle was on the table, and Sheila was inching it closer to the edge. I dared her to tip it over, and she hesitated for only a brief moment. The bottle exploded on impact with the floor, the waiter rushed over, suggested we switch tables while the mess was cleaned, and immediately brought us our food. I was captivated, two-thirds of the way to smitten. And her vibrancy has not abated since then.

ROMANTIC CRUSHES, BAYESIAN STATISTICS, AND LIFE

Romantic crushes can often reshuffle the deck in the card game of life. Alternative metaphor: they can be the joker or the wild card in the biographical deck. Essayist Alain de Botton wrote a blog post titled "On the Madness and Charm of Crushes."[2] As I wrote in the *New York Times*, it deals with a certain sort of crush, a sudden infatuation aroused by the merest of stimuli—the way she subtly rolls her eyes at a blowhard's pronouncements or her intentional dropping of a glass on the floor to attract a waiter's attention (mmm . . .) or the way she casually uses her iPhone as a bookmark.[3] (Pronouns can obviously be switched.) The crush is barely explicable, but is undeniable. In beautiful prose laden with examples de Botton also describes how the attraction often cascades into exultation, but, alas, gradually devolves into disillusionment and a slow vanishing of the mirage.

The article triggered my sometimes reductionist mindset, and I realized that the bare bones of de Botton's thesis could be expressed in statistical terms. To that end let's imagine a person to be an assemblage of traits. Many of these are personal—our looks, habits, backgrounds, attitudes, and so on. Many more are situational, namely, how we behave in the myriad contexts in which we find ourselves.

The first relevant statistical notion is sampling bias. If we want to gauge public feelings about more stringent gun control, we won't get a random sample by asking only people at a shooting range. Likewise, catching a fleeting glimpse of a person or having a brief exchange of repartee with someone elicits a sample of the person's traits so biased that it can, if the situation is right, lead to a crush. But what we find so appealing is often just an idealized caricature; we see what we want to see. In the throes of an incipient romantic fog, we use what psychologist Daniel Kahneman in his book *Thinking, Fast and Slow* calls system 1 thinking, which is fast, automatic, frequent, emotional, stereotypic, and subconscious.[4]

The second relevant statistical notion is Bayes' theorem, a mathematical proposition that tells us how to update our estimates of people, events, and situations in the light of new evidence. A mathematical example: three coins are before you, one is biased, landing on heads only 1/4 of the time, the second one is fair, and the third is a two-headed coin. Pick one of the coins at random. Since there are three coins, the probability that you picked the two-headed one is 1/3. Now flip the one you picked three times. If it comes up heads all three times, you'll likely want to change your estimate of the probability that you chose the two-headed coin. Bayes' theorem tells you how. Specifically it says that the updated probability that you chose the two-headed coin is now about 87.6 percent, up from the initial 33.3 percent.

Obviously the process isn't so cut-and-dried when we're dealing with the convoluted imponderables of personal attraction, but though we're often stubborn about not giving up our illusions, the evidence life presents us eventually forces us to do so. If the crush develops into a relationship, we see the object of our crush in the early morning, doing chores, arguing with us about the myriad of things there are to argue about. Our sample grows more balanced, more representative. Over time we perform in an informal, nonquantitative way what I described above with Bayes' theorem and the three coins. We engage in what Kahneman calls system 2 thinking, which is slow, effortful, infrequent, logical, calculating, and conscious.[5]

Sampling, statistics, and Bayes' theorem are, of course, used constantly in scientific endeavors. There scientists (and mathematicians as well) develop crushes—wild hunches, poorly supported speculations, pet approaches of which they become enamored—on everything under the sun and beyond it. Then they step back and think more slowly and critically to, as best they can, keep from being deluded. Perhaps surprisingly, the dynamics in the two disparate realms do have this point of contact. Whether in abstract science or more generally in everyday life, crushes provide the chaotically bubbling energy of life. We should be skeptical of them, but we should also cherish them.

Needless to add, biographers often develop crushes on their subjects.

DOMESTIC MATH: TOILET SEATS, UP OR DOWN; MOVIES, EARLY OR LATE

Being a mathematician with a longtime interest in applying the discipline to varied and sometimes prosaic issues is part of my defense for including this snippet. The other part is that it is somewhat autobiographical. One reason we read biographies, after all, is to learn trivial, workaday details about a person. It is sometimes satisfying to read that so-and-so puts his pants on one leg at a time (and takes them off the same way). Relevant to this latter is the burning question of whether or not to leave toilet seats in your house up or down.

Let me begin with the premise that raising or lowering the seat is unpleasant—in a very minor sort of way, of course, but unpleasant nonetheless. So, from this it seems to follow that we should endeavor to institute a policy (poolicy, peelicy—I can resist only so many excruciatingly bad puns) that will minimize the number of such toilet seat adjustments. Since men raise the seat when urinating, relevant factors for any proposed policy to take into account should include the relative frequency of urinating versus other uses of the toilet and the number of men and women in the household. Different households will make different assumptions and plug different numbers into a reasonable formula providing an estimate for the average number of times the seat needs adjustment, and the policy in question should be the one that minimizes this number. In most cases, I would guess, the default position should be up, contrary to tradition.

Someone opposed to this probabilistic approach can introduce other desiderata involving, say, neatness or chivalry or "manners." These considerations are less amenable to quantification, and, if pushed hard enough, can win the day against my argument, as indeed they have in my house, alas. Conceding, I've found, is preferable to eliciting my wife's "fine" or "whatever," which terms signify anything but resignation. This matter of toilet seats can even be dressed up as an illustration of the difference between a deontological, rule-based approach to ethics (the default position for the seat absolutely must be down) and a utilitarian, consequence-based approach (let's see what works better for the people in the house). There is value, I repeat, in observing connections between wildly disparate realms, and this example provides an oblique (to say the least) entrée to a large class of ethical dilemmas that often invoke murder cases, pushing fat people over railings, and law courts to illustrate.

Staying on the topic of bathrooms, I recall the common practice of symbolizing simple statements such as "Aristotle had halitosis" or "Thoralf is angry" with the letters P and Q. The practice inspires the only bit of logic-based (public) bathroom humor I know: the difference between men and women is that between the statement [P and not Q] and the statement [Q and not P].

Another crucial domestic issue involves moviegoing. I like to get to movies about five minutes before they begin so I can settle in comfortably and watch the coming attractions. (In this sense only could I be a fan of German philosopher Martin Heidegger's major work whose title *Being and Time* philosopher Sidney Morganbesser facetiously claimed was really "Being on Time.") Sheila likes to arrive two to three nanoseconds early, and this temporal lacuna has characterized our moviegoing for a long time. Since she doesn't wear a watch, years ago I devised a strategy of moving the grandfather clock ahead by a few minutes so that it appeared that we were late and thus should leave earlier. We did leave a little earlier, but after a while she made the natural inference that we had been leaving home too early, and so we could leave later and still get to the movie on time. Thus I was induced to move the clock ahead by a few more minutes. This subterfuge also worked for a bit, suggesting that we were late and should leave earlier. Alas, after a while she again inferred that we had left home too early and could afford to leave later and still reach the movie on time.

After another couple of iterations, the jig was up. Ostensibly we were leaving home a couple of minutes before the movie began and still miraculously arriving a couple of minutes before it began. Now we generally compromise, or if I'm particularly impatient, I walk to the movie a little before she does and save her a seat. Not exactly Zeno's paradox,[6] but it's how this particular tortoise and Achilles make it to the movies. Returning home, we might resume negotiations about the toilet seat.

On to probability and chance, writ small and writ large.

Chapter 8

CHANCES ARE THAT CHANCES ARE

IF ONLY . . . PROBABILITY AND COINCIDENCES, GOOD, BAD, AND UGLY

Most people have a very poor feel for probability, their vocabulary for the notion impoverished, and, for some, it is restricted to variants of these three terms: miraculous, 50-50, and absolutely certain. This indifference to the topic reveals itself in all areas of life, including the way people talk, write, and think about their lives. I sometimes ask students which self-proclaimed psychic is more probabilistically impressive—the one who correctly predicts 54 of 100 coin flips or the one who correctly predicts 26 of 100 coin flips? Most say the first one and don't see how impressively unlikely the second one's predictions are. But anyone who could predict so few correct answers would be invaluable; just negate everything he or she said, and you'd have a true clairvoyant.

This cluelessness about the pervasive mathstuff of chance is accentuated when the events in question are not clear-cut like coin flips but are more nebulous. For example, let's assume something awful occurs, say, a fatal car accident. It's hard not to review the events just prior to the accident and conclude that had any one of them been different, the accident would not have occurred. Furthermore, we can wonder about how probable each of the preceding events was. What if he had bought milk earlier? What if the grocery store had longer hours and he didn't decide to go to the closer convenience store? What if he hadn't taken the shortcut to the store? What if he hadn't made the traffic light and arrived at the next intersection a minute later? What if the driver of the other car had not been fired that morning? What if he hadn't been extended credit at the bar? What if he hadn't been kicked out of the bar at the time he was? And on and on.

We might try to assign a probability to each of these more-or-less independent events. Recall a couple of very useful basic ideas of probability: (i) events are independent if the occurrence of one doesn't affect the probability of the others occurring and (ii) the probability of a number of independent events occurring is obtained by multiplying their individual probabilities. That is, if some event A occurs 1/4th of the time and an independent event B occurs 1/3rd of the time, then B occurs in 1/3rd of those instances in which the event A has occurred, so the events A and B occur together $1/3 \times 1/4$ or 1/12th of the time. Using these ideas, we conclude that the probability of all the accident's antecedents occurring is almost impossibly minuscule.

If, to illustrate, we roll a pair of dice two times, flip a coin three times, and spin a roulette wheel twice, the probability of obtaining a 7, an 11, three heads, and a 27 and 31 is, since the events are independent, equal to the product $6/36 \times 2/36 \times 1/2 \times 1/2 \times 1/2 \times 1/38 \times 1/38$, which equals 1/1,247.616 or about .0000008. This minuscule probability is correct but misleading in the sense that *any* sequence of independent events all occurring will have a vanishingly small probability.

We can, of course, also carry out the above thought experiment for positive events of interest, say, meeting one's future spouse at a protest in Madison, Wisconsin, or having one's book make the *New York Times* bestseller list. What if you hadn't gone to that demonstration or hadn't written that piece for the book review? And if we ask for the probability of the sequence of events leading up to that significant event, the answer will always be that it's tiny, a different tiny number, but still quite tiny. If long enough, any particular sequence of events will be very, very improbable, so the wonderment that is often a consequence of discovering this improbability is natural, but isn't really justified.

Another example: I mentioned earlier that there are almost 10 to the 68th power—a one with 68 zeroes after it—orderings of the 52 cards in a deck. Truly an unfathomable number, but it's not hard to devise even everyday situations that give rise to much larger numbers. Now if we shuffle this deck of cards for a long time and then examine the particular ordering of the cards that happens to result, we would be justified in concluding that the probability of this particular ordering of the cards having occurred is approximately 1 chance in 10 to the 68th power. This certainly qualifies as minusculely minuscule. Still, we would not be justified in concluding that the shuffle's ordering could not have possibly resulted in this particular ordering because its a priori probability is so very, very tiny. Some ordering had to result from the shuffling, and this one did. Incidentally, the vast, vast majority of orderings of the deck that occur during shuffling have never appeared in the history of human card playing.

Resistance to these sorts of considerations stems, I think, from a religious belief that the event in question, whatever it is, with its trail of antecedents was somehow ordained, "meant to be," cosmically special, or a "miracle." If the issue concerns playing cards, most people will buy the probabilistic analysis. Not so with more emotion-laden matters. I've always found it hard to argue with people making such claims with the glaze of certainty in their eyes and their child or a loved one in their thoughts. (More than hard, it's sometimes cruel.)

No one in that frame of mind is here now, so I'll simply make an observation. If an event or sequence of events gives rise to a supposed miracle that is deemed a divine intervention, the one deeming it so must face some obvious questions. Why, for example, do so many people refer to the rescuing of a few children after a destructive tornado as a miracle when they chalk up the death of perhaps dozens of equally innocent children in the same disaster to a meteorological anomaly? It would seem either both are the result of divine intervention or both are a consequence of atmospheric conditions. The same point holds for other tragedies. If a recovery from a disease after a long series of struggles and treatments is considered a miraculous case of divine intervention, then to what do we attribute the contracting of the disease in the first place?

Incidentally, this religious nonsense seems at odds with what the book of Ecclesiastes wisely says: "Time and chance happeneth to them all."[1]

A related factor in our reaction to fortuitous events is how easily imaginable or psychologically available seemingly "nearby" outcomes are. It's easy to imagine missing the traffic light and the accident not occurring. In the other direction, it's also easy to imagine not going to the event and not meeting one's future spouse. The word *nearby* is in quotes because the parallel sequences of events whereby, say, the accident occurs or doesn't occur or you meet or don't meet someone are not necessarily close except in a psychological sense.

Your lottery ticket may differ slightly from the multimillion-dollar winning ticket in that 4 or 5 of the numbers you picked are the same or maybe differ by only 1 from the numbers on the winning ticket. You consider yourself unlucky even though your ticket wasn't close to the winning ticket in any but a psychological sense, which makes winning easier to imagine. Your ticket might as well have had every number numerically distant from the numbers on the winning ticket. There is no reasonable metric with respect to which one ticket is "close" to another or is "almost" the winning ticket.

(I was once on a TV show devoted to science that attempted to make clear how unlikely winning the lottery really is. With cameras running behind me I went into a New York convenience store crowded with ticket buyers, bought a ticket from the clerk, and with a flourish tore it up in front of everyone. You would think that I had kicked a puppy to judge from people's surprise and outrage.)

Finally, I note that this matter of accidental yet life-changing occurrences is, of course, related to the sheer ubiquity of coincidences. Most people don't realize that the number of possible coincidences we might observe grows exponentially with the ever-increasing torrent of numbers, acronyms, facts and factoids, blogs, tweets, ads, and Big Data in general that wash over us every waking moment. The vast, vast majority of this stupendous number of coincidences are meaningless despite the widespread tendency to read personal significance into just about everything.

A standard example of how easily coincidences can arise is provided by the first letters of the planets in order of their distances from the sun and the first letters of the months in calendar order. Is the SUN in MVEMJSUNp significant? (The lower case p reflects Pluto's demotion.) Is the JASON in JFMAMJJASOND significant? Of course not. Neither is the notable (but only notable) biographical fact that Mark Twain was born on the day Halley's comet appeared in 1835 and died on the day it returned in 1910.

Another example is illustrated by the diagram below, a random sequence of 250 simulated coin flips, the Hs or Ts each appearing with probability 1/2. Note the number of runs (strings of consecutive Hs or Ts) and the way there seem to be coincidental clusters and other patterns. If you felt you had to account for these, you would have to invent convoluted explanations that would of necessity be quite false. Random sequences rarely look completely random.

```
THHHT TTHTH HHHTT HHTHH THTTH THTTT TTTHH TTTHH
HTHTH HHHHH HHHTH HHTHT HHTTT HTTHH HTTTH HHHHT
THHTT THHTT TTTTH HHHTH HHHHTT THHHT TTHTH HTTHH
THTHT THHHT HHTHH HHTHH TTHHT HHHHH HHTHT HHHHT
TTTTT HTTHH HTTHH HHTTH HHTHH HHTTT HHTHH HTTHH
TTTTT HHHHT TTTHH HHTHH TTHHH HHHTH HTTTT TTटHT
HHHHH TTTHT
```

Finally, let me illustrate with a more extended instance I've written about before. It involves 9-11, a numerical phrase that is happily and finally appearing less and less frequently. On Wednesday, September 11, 2002—9/11/02—the New York State lottery numbers were 911, an eerie coincidence that set many people to thinking or, perhaps more accurately, to not thinking. A natural question comes to mind: How likely is this? After all, the lottery took place in New York State on the anniversary of the mass murder exactly one year before. Remember, however, that on any given day, each of the 1,000 possibilities—000, 001, ..., 233, ..., 714, ..., 998, 999—is as likely to come up as any other. This is true of September 11 as well, so the probability that 911 would come up on that date is simply 1 in a 1,000. This probability is small, but not minuscule.[2]

This leads to what I like to call the Fundamental Confusion of Coincidences: the probability of an unusual event or sequence of events is usually very small, in fact often minuscule, yet the probability of some event or sequence of events of the same vaguely defined general sort is usually quite high. People regularly confuse these two probabilities.

Thus the broader, more appropriate question that should come to mind regarding this 911 story is: What is the probability that some event of this same general sort—something that is resonant with the date or is likely to stimulate us to think of it—would occur on September 11? The answer is impossible to say with any precision, but it is quite high. First off, there are two daily drawings in the New York State lottery, so there were two chances for 911 to come up that day, increasing the probability to (a bit under) 1 in 500. More importantly, there were innumerable other ways for an uncanny coincidence to occur.

How many addresses or license plates, for example, have 911 in them? At each of these addresses and for each of these vehicles, something could have occurred that caused people to think of September 11. Possibilities include an accident, murder, or arrest of someone suspected of terrorism, related to a victim of the attack, or otherwise associated with it. Or consider sports scores and statistics. The stock market is also a major producer of numbers. Another "close" example is the September 10 closing value of the September S&P 500 futures contracts. It was 911.

But this is all too easy to do. There are an indeterminate number of ways for such events to come about even though the probability of any particular one of them is tiny. Furthermore, after such an event occurs, people glom onto its tiny probability and, falling victim to the Fundamental Confusion, neglect to ask the much more pertinent question: How likely is something vaguely similar to this to occur?

Keep this in mind when you read the following excerpt from the great science fiction author Arthur C. Clarke. In his 1973 novel, *Rendezvous with Rama*, Clarke wrote: "At 0940 GMT on the morning of September 11 in the exceptionally beautiful summer of the year 2077, most of the inhabitants of Europe saw a dazzling fireball. . . . Somewhere above Austria it began to disintegrate. . . . The cities of Padua and Verona were wiped from the face of the earth, and the last glories of Venice sank forever."[3]

Similar coincidences occur in our personal lives and biographies, where their insignificance may be even harder to acknowledge because of our familiarity with everyday situations. Our always active pattern-seeking brains lead us to see meaning and an agent behind almost everything. This may be a consequence of our evolutionary development in information-poor environments, but whatever the cause our brains have been primed to see patterns that aren't there, to infer agency where there is only randomness, and to invent nonexistent, seemingly explanatory entities. Sadly, one very important realm in which the assertion of randomness is false is natural selection, which many creationists continue to proclaim is a completely random process.

As I've noted elsewhere, surely the most amazing coincidence of all would be the complete absence of all coincidences.

INNUMERACY, *A MATHEMATICIAN READS THE NEWSPAPER,* AND THEIR AFTERMATH

For a while after coming to Temple University from the University of Wisconsin I taught an outside course on basic mathematics, mainly arithmetic, to nurses at a local medical school for a little extra money. The people in my class—mostly women, a couple of men—seemed temperamentally suited to the profession, appearing to be compassionate, empathetic, and caring. Unfortunately most of them were quite innumerate. Try as I did to get across the rudiments of percentages, numerical prefixes, proportions, simple calculations, and unit conversions, many never managed to distinguish 2 percent from .02 percent nor could they, despite my pleading, believe it was even very important to do so. The following semester I was scheduled to teach the course again and told the program supervisor that this level of mathematical naïveté was dangerous and that I wouldn't want to be attended to by many of the soon-to-be nurses were I to be hospitalized. The next day I received a letter informing me that my services would no longer be needed.

Although I'd been aware of widespread innumeracy for a long time, this incident and the general dismissal of the relevance and importance of mathematics, even by some very accomplished people in organizations (including writers' groups) to which I belonged, were exasperating and catalyzing. I wrote a piece on the subject for *Newsweek*, and at the behest of Rafe Sagalyn, who noted it and soon became my agent, I expanded it into my book *Innumeracy*, which was published by Farrar, Straus and Giroux in 1989. A book on the mathematical aspects of various topical issues and the consequences of mathematical illiteracy, it became a national bestseller that remained on the *New York Times* bestseller list for four and a half months.[4] Although I'd had the experiences related above and others of the same type, its success was a surprise and something of a turning point in my and my family's life. I don't want to be falsely modest; it is a very good book that promulgated an important message, imparted some essential mathematical ideas, and popularized a most useful word (humble brag). I was surprised at the wonderful reviews the book received, among them one by the aforementioned Douglas Hofstadter, author of *Gödel, Escher, and Bach*, who enthused, "Our society would be unimaginably different if the average person truly understood the ideas in this marvelous and important book."[5]

Still, an important point I'd like to emphasize is that, as with many such turning points, a huge element of luck was involved. Probably essential to the book's success is that the date of publication just happened to coincide with the release of the first widely heralded report on the mathematical failings of American students. This news hook gave warrant to people wishing to mention the book in places and contexts other than the book section. Besides the *Newsweek* piece mentioned above, I had also recently written a piece on innumeracy and literature for the *New York Times*, and these two articles led to an appearance on NBC's *Today* show, which would not have occurred except for a last-minute cancellation by another guest. After these fortuitous events, I witnessed the exhilarating effects of the media's hall of mirrors, whereby one appearance leads to another, which spawns several more. No online social media or blogs existed in those antediluvian times, so the appearances were on local television and radio across the country and, as the book became better known, in the national media as well, from NPR and talks at the Smithsonian, the National Academy of Sciences, and Harvard's Hasty Pudding Club, to *Larry King Live*, *People* magazine, and *Late Show with David Letterman*. The book helped jump-start a new wave of discussions about mathematical pedagogy (culminating in the so-called Common Core and a slew of other STEM-related initiatives stressing a more conceptual understanding of mathematics) and was even the answer to a question on the popular game show *Jeopardy*.

One appearance was particularly memorable. I was being interviewed on a local morning show in Texas by a former beauty queen who seemed miffed that she had been dragooned into interviewing a mathematician with unruly hair. Her assistant stood behind me holding up posters from which she read the questions in a singsong, perfunctory way, her eyes glazing over until the next poster went up behind me. Exhausted from the book tour and a bit annoyed, I waited until the assistant put the poster down and asked the interviewer to repeat the question. As I suspected she wasn't even aware what she had asked me and stumbled badly. After the shortened interview she hurried off the set, and the segment producer thanked me for underlining what she had been telling the station management for months.

The whole experience was fun. I had caught a wave and felt like a passive passenger on a jerky Ferris wheel as the interviews, profiles, and speaking invitations poured in. The publicity has, of course, subsided, but it led to a change in my self-conception from mathematician to writer. This transition was facilitated by the aforementioned fact that my educational background at Wisconsin was a bit anomalous for a mathematician. My choice of a possible undergraduate major changed from classics and English to philosophy and physics and ultimately to mathematics. I have remained throughout a mathematics professor at Temple University, a state university in Philadelphia, where I still enjoy teaching a wide variety of courses to a wide variety of students. I have, however, devoted much of my time to writing eight (this is the ninth) books ranging from *A Mathematician Reads the Newspaper*[6] to *I Think, Therefore I Laugh*,[7] as well as columns, articles, and op-eds for the *Guardian* (in the United Kingdom), the *New York Times*, and a host of other publications on issues ranging from mammogram false-positives to the mathematics of the BP oil spill. For ten years I also wrote the monthly "Who's Counting" column for ABCNews.com.[8] It dealt with a variety of mathematical issues, especially mathematical aspects of stories in the news, and was a task that greatly increased my respect for columnists who must write something topical and insightful two or three times a week.

A Mathematician Reads the Newspaper was published a few years after *Innumeracy*, and it also fared quite well, even making the Random House list of the 100 best nonfiction books of the century. (I've resisted my natural tendency to debunk the list.) Because of the book, *Innumeracy*, my columns, and other writings I was asked by the dean of the Columbia School of Journalism to develop and teach a course on quantitative literacy for aspiring journalists. I did so but was disappointed, haunted even, at the mathematical naïveté of the students in the course, who made it necessary to begin with remedial arithmetic instead of the problems and odd-ities associated with numbers in the news. Such a course should be a requirement at Columbia and at all schools of journalism, but unfortunately it is not.

Another consequence of the newspaper book was that I was asked to write weekly op-eds and editorials for the *Philadelphia Daily News*. I enjoyed writing them, but the letters to the editor the paper received in response to some of them were a little dispiriting, at least at first. I felt a little like Nathanael West's anonymous male columnist Miss Lonelyhearts, although the title Miss Emptyheads would have been more apt. After a short while, however, I found the letters and misunderstandings humorous in something like the way I still find it amusing to give the name Ludwig Wittgenstein (or Baruch Spinoza or René Descartes) to snooty restaurant maître d's and later hear "table for Ludwig Wittgenstein, right this way."

As a writer on many topics, most but not all tangentially related to mathematics, I've met a number of interesting people whom I likely would not have met otherwise had I not been lucky enough to receive the publicity and recognition I did. A minor drawback associated with my writings, however, is that I was, and to an extent still am, asked the same questions repeatedly. For a short while the question was: What is David Letterman really like? Answer: How am I supposed to know? Letterman seemed like a cool professional with a quick wit and a riveting gaze.

Perennial questions are: Is there anything new in math to

be discovered? What can you do if math was your worst subject and you don't have that kind of brain? How long did it take you to write this or that book? What is the probability of this or that coincidence? (Allow me to indulge once again my professorial tendency to reiterate and note that though the probability may be quite small, it is very often the answer to the wrong question. The right question is usually: What is the probability of something of this vaguely defined general sort occurring? and This probability is almost always much larger.)

The most common pedagogical question is and was: Should students be able to use calculators, web apps, and related aids? Answer: The short one is yes. The question, however, suggests a very narrow conception of mathematics as essentially glorified computation. The truth is that computational skills aren't much more essential to mathematical understanding and modeling than typing skills are to reading and writing. As I've stressed repeatedly, mathematics is no more computation than literature is typing. I repeat: mathematics is no more computation than literature is typing. (And nobody says you're a slow typist, so you had better forget that novel you're working on.) Often what is crucial in both journalism and mathematical applications is context, a bogus etymology for which is *conte*, the French word for *story*, and x and t, the most common variable names used in mathematics.

A somewhat darker aspect of writing, at least for me, was the many vituperative letters and e-mails I received over the years because of my writings, especially because of *Innumeracy*, in which I made a few innocuous comments, one being that I always use my middle name in public contexts to distinguish myself from the then pope, John Paul. My book *Irreligion*,[9] although not particularly snarky, and my columns have also elicited hate mail. Simple multiplication offers a partial explanation. Even if the percentage of very angry, deranged, or self-righteous people is tiny, if your work reaches a large enough number of people, the product of these two numbers will generate a fair amount of nasty and nutty stuff: self-proclaimed

Nazis denouncing this or that, pictures of me with a bull's-eye on my forehead, schizophrenic ramblings, and much more.

A minor, less alarming downside to the publicity my writings have received is that I'm often termed a "crusader" for mathematics. I am flattered by the metaphorical use of the term, but it also makes me wince a little. Mathematics doesn't need any human crusaders; not only is it a beautiful subject and critical for the development of science and technology, but it is also an imperialist discipline that, broadly interpreted, can invade and occupy almost every other discipline, arguably including the writing of biographies. To demonstrate its all-encompassing range, I've added to whatever mathematics course I have taught over the years odd examples, seeming paradoxes, everyday applications, probability and logic puzzles, and open-ended problems. I've also made liberal use of a bit of pedagogical wisdom suggested by April Fools' hoaxes: in class I sometimes intentionally talk BS to encourage skepticism and critical thinking. I hate students' mindless nodding at whatever it is Herr (or in my case Hair) Professor says, no matter how nonsensical.

This approach and much of the material, usually in the form of little vignettes rather than formulas or equations, have made it into my books and columns. Still, exploring the ever-increasing salience of mathematics, expounding some mathematical notions that belong in everyone's intellectual tool kit, and pointing out the consequences of innumeracy does not make me a "math crusader" brutally converting the innumerate infidels but rather simply a mathematician who writes. I do, however, take a tiny smidgen of credit for the now larger number of very talented authors devoted to communicating mathematical ideas and applications.

Finally, the crusader characterization does suggest a complexity-theoretic point: being typecast is unavoidable. Necessarily we all are to one degree or another, and the reason is that we're only capable of handling a limited amount of complexity and nuance, and labels and stereotypes are often useful ways of summarizing data sets or people that are beyond our complexity horizons. The bottom line is that I'm that still rather odd creature, a mathematician who writes, and therefore I'm perhaps a bit like a dog who plays chess. What's remarkable is not that the dog isn't very good at the game but that he plays at all.

Chapter 9

LIVES IN THE ERA OF NUMBERS AND NETWORKS

HOW MANY E-MAILS, WHERE DID WE BUY THAT— THE QUANTIFIED LIFE

I've never gotten over my puerile pleasure in counting items, including those of only personal interest. I've been in forty-eight states and forty-four countries; I usually keep track of the number of miles I've driven on the New Jersey Turnpike when driving to New York; I know the number of steps from the basement up to my office; I know the approximate number of e-mails I receive daily; I know how many Twitter followers I have, and so on. I won't list 293 more examples here, since it's enough to say that I possess an almost embarrassingly large cache of numerical selfies in my relatively large head.

Stephen Wolfram does me one better (or rather X times better). He is a computer scientist and the creator of Mathematica software and the online answer engine Wolfram Alpha, as well as the author of *A New Kind of Science*. In 2012 he published on his blog a summary of the self-tracking he'd done utilizing various bits of software, mostly written by him. He tracked the number and times of incoming and outgoing phone calls he'd made over the previous twenty or so years and graphed their hourly, daily, and monthly distributions. He did the same for his e-mails, as well for the number and distribution of his engagements, events of one sort or other, and conversations over the same period. He even recorded the number of keystrokes he tapped out on his computers as well as the number of steps his pedometer indicated he'd walked. In short, he did for himself what the Internet and corporate data collection seem to be doing for all of us.[1]

All of these quantities were recorded passively and graphed for ease of interpretation by the software. All in all, the exercise provides a fascinating portrait of parts of Wolfram's life. Moreover, this idea of the "quantified self" has spread and will continue to do so, no doubt supplemented by ever more pervasive social media that will allow people to construct ever fuller portraits of their own lives (I'm tempted to say their own several lives. [Recently Seth Stephens-Davidowitz wrote a piece in the *New York Times* contrasting the numbers people report in reputable surveys about their sex life and the generally much lower numbers that Google searches and simple arithmetic strongly suggest.[2]]) With such a record we would be able to note what time of day was most productive, what seasons were busiest, and at what rate projects were completed. Wolfram, for example, recorded at what times he worked on different chapters of his book and when he modified various of his voluminous computer files.[3]

Wolfram also compiled data on Facebook subscribers and used it to focus on age and, by extension, biography. He looked at the age of Facebook users and their relationship status, the number of their friends, their friends' median age, the number of friend clusters, their connectedness, and the topics discussed. He focused, to use his words, a serious "computational telescope" on the "social universe."[4]

Despite my penchant for counting, I never came close to amassing this much numerical data about myself and wish I'd had access to such data about me for the last couple of decades. But maybe not. In my case I suspect they'd reveal fallow stretches during which I wasn't doing much of anything except possibly taking a lot of pedometer-recorded steps to retrieve Diet Cokes in order to refresh myself during my desultory surfing. Maybe with the benefit of this social telescope I would have seen that such behavior was common.

Still, more important than the numbers of e-mails, computer files, calls, etcetera is their content—the data, not just the metadata. I recently went through some old letters and photos and realized that my memories of various events were cloudy at best. Looking at old home movies has the same effect; I forgot that I acted like a doofus in many of them. My memory of physical items is even worse. In fact, another possible dimension to a quantified life might involve one's purchases and possessions, both prosaic and special. Regarding the latter, my wife sometimes points to some item in the house and asks for its backstory. Where did we buy it? When did we do so? How much did it cost? Or did somebody give it to us? And if so, why? I usually fail abysmally at this game, but I would guess, perhaps wrongly and a little defensively, that most mathematicians have a similar indifference to possessions. Sheila excels at this game but notes sarcastically that I do know important things like who succeeded John Foster Dulles as secretary of state under Dwight D. Eisenhower (Christian Herter). I might do a little better at this item-identification game if our credit card expenditures were fed directly into a Wolfram-style website that I could check on my smartphone.

Of perhaps some relevance is a bit of new neurological evidence that memoir and math are somewhat incompatible activities. Josef Parvizi and other Stanford University researchers published an article in the *Proceedings of the National Academy of Sciences* in which they concluded, "Particular clusters of nerve cells in the PMC (posterior medial cortex of the brain) that are most active when you are recalling details of your own past are suppressed when you are performing mathematical calculations."[5] This incompatibility shouldn't be too surprising; one activity involves the specific and personal, the other the general and abstract. It is, in fact, this chasm I'm trying to bridge in this book. Interesting question: Are people's memories of their youthful exploits as innumerate as they sometimes seem?

In any case, it's so easy to forget that relationships, incidents, and purchases were not as remembered, and documents and the numbers in them keep us from projecting present circumstances and views backward into the past. Without some sort of documentation we tend to select the pages and (often false) details of our life story most consistent with whatever grand narrative we like to tell about ourselves. I'm sure I'm somewhat guilty of that here.

Incidentally, acknowledgment of this foible is a version of the so-called paradox of the preface, wherein an author both acknowledges mistakes and falsehoods in his or her book but nevertheless supports every individual statement in it. In other words, you can always think you're right, but you shouldn't think you're always right. (The placement of "always" like that of "only" can drastically change the meaning of a sentence. It doesn't always do so, only sometimes.) In the related lottery paradox, one believes that each individual ticket is a losing ticket, yet also believes there is a winning ticket.

Of course, it was never Wolfram's intention that his quantified self would be a real biography or even a brief memoir. It's not, and what it leaves out may well be more biographically important than what it includes. Imagine, for an extreme example, French novelist Marcel Proust and *In Search of Lost Time*. The book is ruminative, endlessly reflective, and attempts to plumb remembrances quite different from those measured by Wolfram. Although it would be interesting to see how Proust worked through the night, the distribution of different words and phrases in his work, and the number of steps he took pacing around his cork-lined bedroom, this data wouldn't help much in understanding his life or his work.[6]

Do we want amply realized people, even fictional ones, or numbers and demographics—either Leo Tolstoy's *Anna Karenina* or Saul Bellow's *Herzog* (one of my favorite novels) on the one hand or, on the other, statistical factoids such as that the average resident of Dade County, Florida, is born Hispanic and dies Jewish? Artist Pablo Picasso's remark comes to mind: "Art is a lie that makes us realize the truth."[7] But documents, both their number and content, help us see it too.

A TWITTERISH APPROACH TO BIOGRAPHY

Another approach to biography is to focus on the subject's connections to others, rather than on his (or her, throughout) documents, activities, and psychology alone. What was the subject's status within his family, within his fame-inducing sphere of influence, within his immediate physical environment? The impact of sociological, historical, game-theoretic, and other disciplines of relevance are beyond the scope of this book, but they do suggest, as does much else, that the notion of an individual's biography is problematic. Any attempt to fully tell a life story requires an undertaking of colossal proportions. (Robert Caro's five-volume—the fifth not yet finished—biography of Lyndon Johnson comes as close as any, but even it falls far short.)

Rather than wade into these very deep waters, however, we can zoom in on a simpler, quantitative picture of a subject's connections to others: contemporary social media. People's online social networks and their evolving positions within them—constituting a trajectory of sorts along various dimensions—can't reveal what the disciplines mentioned above might, but they do tell us much about a subject that a myopic focus on them as isolated individuals does not. In fact, social media have already subtly reshaped our notion of personal identity.

Consider, for example, the social network, Twitter, to which I belong, where at the moment I'm listed as one of the top fifty science "stars."(The distancing quotation marks are mine.) Like all networks it can be considered a collection of points (people, in this case) linked by lines (the relation of following or being followed, in this case). The collection of points changes as new people enter and leave Twitter, and the lines linking you to others also changes as people follow or unfollow (an awful but common verb in this context) you, and you follow or unfollow others.

My particular experience with Twitter largely involves commenting (insightfully, I'd like to think) on topical issues, whether they be mathematical, political, or broadly cultural, and/or remarking upon and linking to online articles that I find appealing. I do this via short tweets of fewer than 140 characters, which obviously puts a high premium on pithiness. Anyone who tweets realizes that this pithiness poses a significant difficulty that more discursive writing does not. Recall the quote attributed to seventeenth-century mathematician Blaise Pascal, who famously wrote, "I would have written a shorter letter, but I did not have the time."[8] (Variants of the quote have also been attributed to Mark Twain, among others.)

(Incidentally, these 140 characters allow for a staggeringly astronomical number of possible tweets. If you count uppercase and lowercase letters, numerals, special symbols, brackets, and punctuation marks, there are about 100 possible symbols for each of the allowed 140 characters. Thus, there are approximately 100^{140} possible tweets or, after performance of a little arithmetical manipulation, googol$^{2.8}$ possible tweets, where a googol, from which Google derives its name, is 10^{100}, one followed by 100 zeroes. Of course, almost all of these tweets will be completely nonsensical as, unfortunately, are a significant fraction of the ones that make superficial sense.)

Long before Twitter, I was a firm believer in the value of succinctness, especially in mathematical writing. I sometimes perform a complicated magic trick for my classes, repeat it slowly several times, reveal its secrets, and then ask them to write an expository piece explaining it. They sometimes complain that it is unfair to require this of math majors and particularly resent my grading policy. Their grades, I announce to them, will be inversely proportional to the number of words in their pieces provided they at least capture every salient detail of the trick. I tell them that a sense of elegance and concision is essential in mathematics, and this is one way to nurture this sense, but not too many buy it. Whether for my pithiness, which is not particularly tweet-worthy in this paragraph, or for other reasons, the number of people who follow me has grown and changed considerably and is much larger than the number that I follow, but I learn of topics, issues, and articles from those I do follow that I often wouldn't otherwise come across.

Moreover, a simple list of my tweets over a period of time provides a useful reminder of my passing interests, a kind of mini-autobiography of the period. I grant that it is about as evanescent as a raindrop on a windblown piece of litter, but so in a way are the everyday moments of our lives, which are no less valuable for their evanescence. As I tweeted once, "Heraclitus understood the essence of twitter 2500 years ago: 'One can't step in the same river twice. Everything flows and nothing stays.'"[9]

I'm often surprised by the range of people who retweet me (send out my tweets to their followers). Several quotations from my writings regularly ricochet through the far reaches of the twitterverse, including, for example, "The only certainty is uncertainty, and the only security is learning to how to live with insecurity." I shouldn't be surprised at this, as the so-called Watts–Strogatz model of such networks suggests as much. The model does not link points at random. Nearby (in one sense or another) points are more likely to establish a link leading to clustering (a high density of links within a small group), but with an appropriate small admixture of links between more distant points. This latter feature allows the average path length between any two points in the network to be relatively small. One needn't know more than a few people in a distant country, for example, to be somewhat closely linked to most people in that country. The clustering and small-world properties of such networks are typical of many social networks, whether online or elsewhere.

The path length between two points leads to the notion of six degrees of separation between people in real-world social networks. The idea is that most people are connected via six or fewer links, the average number varying depending on the specific network. (In the United States the average number is probably less than six.) There are points—people, that is—that can be considered central hubs since they have links to far-flung other points. These people are often of great interest, and many subcultures have their own version of a natural hub for such linkages. Most know, for example, that the game "Six Degrees of Kevin Bacon" refers to the number of movie links between an arbitrary actor and the actor Kevin Bacon. An actor is directly linked with Bacon if they both appeared in the same movie and indirectly linked if Bacon and (s)he both appeared in movies with the same third actor or, more commonly, via a set of more numerous intervening movie links.

The Kevin Bacon game is related to the notion of an Erdős number among mathematicians. Paul Erdős was a prolific and peripatetic mathematician who coauthored more than fifteen hundred papers, many with a diverse set of mathematicians around the world. One's Erdős number (0 for Erdős, 1 for a coauthor with Erdős, 2 for someone who coauthored a paper with a coauthor of Erdős, and so on) is a measure of the distance between a mathematician and Erdős. That is, two mathematicians are directly linked if they were among the coauthors of a paper in the same way that two actors are directly linked if they appeared in the same movie.[10]

My Erdős number is 4, largely due to a couple of quite fortuitous links. I also had a personal link to this same man, Erdős, who scared me witless one night around three a.m. I was in graduate school at Wisconsin and was working late in my office. Certain that no one else was in the building, I wandered through the halls barefoot as I contemplated whatever it was I was contemplating. Behind me a little voice asked what I was working on. I turned and jumped but managed to keep from yelping in terror. It was Erdős, and, most flatteringly, he really did want to know what I was working on. We discussed it for a while, and he proceeded on his way with his signature cup of coffee in his hands. One of his well-known quips is that a mathematician is a machine for turning coffee into theorems.

(Rather parasitically, I sometimes say that I turn Diet Coke into paragraphs.)

An alternative approach to small networks, say, students in a class or relatives at a reunion, is a so-called incidence matrix, a rectangular array of numbers, each a 0 or a 1, indicating whether people are connected or not. If there were twenty people, for example, there would be twenty rows of twenty numbers, the number in the ith row and jth column being 1 if person i is connected to—let's say can initiate contact with—person j and 0 if not. The presence of a 1 in the ith row and jth column does not always imply a 1 in the jth row and ith column. (Don't call me; I'll call you.) Various mathematical techniques can then be applied to the matrix to yield further information. By multiplying this matrix by itself in the special way in which matrices are multiplied, we can easily determine the number of two- and three-step communication paths from i to j and the identity of central figures or hubs in the group. One could also infer the existence of cliques and dominance relations in the group. Constructing such an incidence matrix for one's friends or relatives is instructive, but it's probably wise to keep your assignment of 1s and 0s secret.

However it's modeled, the notion of a network is quite flexible, and we can certainly stretch it to encompass our personal mental network, arguably what our "self" is. It is constituted by emotionally important people we've known, some vivid incidents in our lives, geographical landmarks, evocative smells, songs, and words or phrases, and other hub-like points. These people, events, places, smells, songs, words are especially resonant because they connect to and activate so many others. They're the Kevin Bacons and Paul Erdőses of our private lives. There is undoubtedly a corresponding neural network in what novelist David Foster Wallace terms our "2.8 pounds of electrified pâté."[11]

A few of my personal mental hubs in no particular order but all capable of delivering pinpricks of nostalgia and intimations of mortality are the Uptown Theater (where I sat through innumerable B movies as a kid); Burleigh (a big street near my childhood home);

my immediate family, of course, various old friends, colleagues, and relatives; Limekiln Pike (near our first home outside Philadelphia); pastitsio as well as cinnamon on burnt butter spaghetti (my grandmother's dishes); Wilbur Wright (my junior high school); Dodo (an early nickname for my brother Jim); Alex's popcorn truck (a family friend's livelihood); Colfax (a street in Denver near which I spent summers); Leonard Cohen's "Suzanne"; piffle (my father's refrain); Wachman Hall (location of Temple University's math department); Riverdale (where my wife's parents lived), Kakamega (the secondary school in Kenya where I briefly taught); Ozzie and Harriet; Buddy Holly; resonant pop music; the New Jersey Turnpike; Van Vleck Hall (the tall math department building at the University of Wisconsin); and the dirt alley behind my grandparents' home.

Expanding a little on the latter hub, for example, I remember that this alley was always infested with hundreds of maniacal grasshoppers that seemed to scatter drunkenly as we walked through the alley on the way to Colfax Avenue where the outside world began. The alley also brings to mind the drugstore on Ivanhoe where I used to sneak peeks at *Playboy* (it was a different time); King Super's and Save-A-Nickel; the Chat and Chew restaurant; the newspaper box on Kierney where I'd pick up the *Rocky Mountain News* in the morning; and, more generally, the idyllic, faraway Denver of the 1950s.

There are simple properties of networks, whether virtual or real, personal or abstract, which are often surprising. For example, are your friends on average more popular than you are? There doesn't seem to be any obvious reason to suppose this to be true, but mathematically it is. As sociologist Scott Feld first observed, the key to the mathematical proof is that we are all more likely to become friends with someone who has a lot of friends than we are to befriend someone with only a few friends. It's not that we avoid those with few friends; rather it's more probable that we will be among a popular person's friends simply because he or she has a larger number of them.[12] This simple realization is relevant not only to networks of real-life friends but also to social-media networks. On Twitter and Facebook, for example, it gives rise to what might be called the Friend or follower paradox: most people have fewer Friends and followers than their Friends or followers do on average. Before you resolve to become more scintillating, however, remember that most people are in similar, sparsely populated boats.

Whether one is popular or not, examining one's evolution, connectedness, and trajectory through social networks affords a sort of external biography of a subject. It's more inclusive, but not that different from tracing a person's rise (or fall) in a corporate hierarchy. Considering a person's position in a network is also a helpful corrective to the belief that a person's traits or attributes more or less determine his or her decisions and actions. More often the person's situation and position in an appropriate network are more determinative, and critical pundits on the sidelines (such as myself) would likely behave in the same way as those they're criticizing were they similarly situated.

Another surprising property is that one's position in a network is constraining, not only in an obvious social sense but also in a topological/geometrical one. The eighteenth-century Swiss mathematician Leonhard Euler famously proved that crossing all seven bridges in the town of Königsberg exactly once and ending up where one started was impossible. The town straddled a river that contained two islands, and Euler realized that the bridges, landmasses on either side of the river, and the islands could be viewed as a simple network. The impossibility of a circuit over all the bridges led to perhaps the first theorem of modern network theory, but because of the pervasiveness of Big Data, the ever-increasing amount of data in social networks, or even in our own psychological networks, the field is now a burgeoning one. Someday we'll perhaps be able to use as yet undiscovered topological/geometric theorems about networks to better describe the higher-dimensional "shape" of our lives and to better explain, among other things, why some seemingly natural paths and circuits are impossible. Of course, we already know without the benefit of network theory that we can't all be baseball players or comedians as well as mathematicians. None of us can cross every bridge we want to.

Finally, I note that the impact of our growing interconnectedness on social-networking sites and the ever-expanding prevalence of smartphones, tablets, and other devices is impossible to predict. Seeing young people in India, Indonesia, Morocco, Peru, here, and elsewhere texting away or checking their Twitter and Facebook accounts suggests that the bonds to their ambient cultures will soon weaken. One guess I will hazard is that social-networking sites and the Internet will make it increasingly difficult for great hero or great villain status to be conferred upon people. The alleged saintliness or deviltry will be too easy to demystify and debunk. A torrent of tweets can be surprisingly powerful.

SCALE AND PREDICTABILITY

As with our position in various networks, sheer size in various senses, especially the physical, also plays an important but sometimes invisible role in how we view the world and ourselves. We all start out small, a fact that's always struck me as underappreciated, and this perspective probably stays with us forever even if we grow up to be very large people. Responding intelligently to changes in size and to nonlinear scaling up or down is not something that comes easily to people even in purely physical situations. Given the choice, for example, of buying three meatballs each 3 inches in diameter or fifty meatballs each 1 inch in diameter for the same price, most will choose the fifty meatballs, even though the three large meatballs provide 60 percent more meat. (The volume of meatballs scales up with the cube of their diameter.) More common are similar misjudgments about the relative value of small and large pizzas. I've often seen people order two expensive 8-inch pizzas, for example, rather than one proportionally much less expensive 14-inch pizza even though the latter is three times the area of one of the former.

(Geometry question: what is the volume of a pizza—geometrically a thin cylinder—of radius Z and thickness A? Answer: PI*Z*Z*A, or PIZZA. By contrast, the volume of a deep-dish pizza of radius L and depth O is PI*L*L*O, or PILLO.)

These considerations also explain why there could never be a King Kong, a gorilla ten times the height of a normal gorilla but proportionally shaped and made out of the same "gorilla stuff." If such a super gorilla were ten times as tall as a normal one, it would be 1,000 or 10^3 times the weight, since weight, like volume, scales up with the cube of the scaling factor. Thus if a normal gorilla weighs 400 pounds, King Kong would weigh 400,000 pounds, much too much for even the enlarged cross-section of its legs and spine to support. King Kong would need immediate hip and knee replacements.

The same sorts of size and scaling difficulties but of a vastly larger order characterize governmental policies and even personal decisions every day, ever more so now that terabytes of data can be stored on devices smaller than one's hand. The policies that work in Rhode Island may not work in California, and the analysis that helps explain one person's behavior may not help much with a person of a different psychological size. Like meatballs, most entities of interest do not scale up in a nice linear manner. Taking account of the effect of scale is vastly more difficult when the dimensions involved are not only physical or informational but less quantifiable as well. How, for example, do we view our actions and their consequences—from psychologically up close or from a lofty disinterested perch, on some absolute scale or relative to social context?

Scale and size are, of course, important in biography and its impact. Consider a rather strained but fanciful analogy. What can happen when a private life story becomes public property is that the audience for the story, limited and flawed as the story is sure to be, expands so fast as to distort the image of the person and render parts of his or her life invisible. The occasional kindness of horrible people gets lost, as does the occasional cruelty of wonderful people. The far-fetched analogy: Think of the inflationary universe hypothesis that holds that after the big bang, a tiny bubble of the primordial universe inflated so fast that parts of the universe have become invisible. The simple point is that a widely known biography can make the subject's private life almost disappear.

More everyday examples arise when we, biographers in particular, try to decide whether certain kinds of decisions and actions in a person's life count as important or not. This is often a matter of how we perceive relative size and scale. Was it a little skirmish over some trivial remark or a pivotal battle about core beliefs? Do we consider the expenditure for the subzero refrigerator to be a small part of the cost of an expensive new kitchen, or do we consider it in isolation and think it to be exorbitantly large? With regard to the latter, I note the relevance of the pervasive psychological foible known as the anchoring effect. It occurs because people are unduly influenced by, or anchored to, some initially presented number or bit of information, whether reasonable or not. (If you ask people, for example, to very quickly guess the value of 10!—read 10 factorial—and define it to be $10 \times 9 \times 8 \ldots 3 \times 2 \times 1$, their guesses will be higher than if you define 10! equivalently as $1 \times 2 \times 3 \ldots 8 \times 9 \times 10$, presumably because people become anchored to the initial 10 rather than to the initial 1.)

Proust did not know the term "anchoring effect," but he noted its relevance to memory, nostalgia, and biographical description. If at a reunion after forty or fifty years, for example, you begin with the assumption that your old friends and classmates have remained as you remember them, you will find them to have aged considerably. You're anchored to their early looks. On the other hand, if you begin with the assumption that they've no doubt aged considerably, you will soon find them looking remarkably like their former selves. In this second case, you're anchored to their imagined aged looks.

Anchoring and nonlinear scaling impact individual lives in different ways (as do meatballs). The same benefit conferred on two people may help one thrive as a kind of personal multiplier effect kicks in and do nothing for the other. Nonlinear scaling can be much more problematic than not getting a good buy on purchases, however. When nonlinear equations feedback on each other, they can give rise to chaotic dynamics, a characteristic of which is the sensitive dependence of a phenomenon on initial conditions and the earlier mentioned butterfly effect, making precise prediction in these situations all but impossible. Since tiny variations in initial conditions can cascade into quite different weather phenomena, weather prediction that is both precise and long-term is impossible. Our psychologies, like the weather and the earlier discussed pinball machines, are complex nonlinear systems, and just as sudden storms develop unpredictably, so sometimes do dark moods overtake us suddenly and unpredictably. This holds for sometimes inexplicable but joyous sunny moods as well or, as Russian writer Alexander Herzen describes them in his memoir, "the summer lightning of individual happiness."[13]

In any case whether a phenomenon of any sort (chaotic, nonlinear, or simple linear) appears deterministic or probabilistic is sometimes a matter of the scale and perspective we adopt. Take shooting pool, for example. It is generally considered a deterministic process. Assume that the angle of incidence equals the angle of reflection, factor in spin, and everything follows. If you manage to hit a ball in the *exactly* correct direction and spin, it will hit another ball, drop into the pocket, or do whatever it was you intended it to do. Contrast this with the flip of a coin, which is usually seen as a probabilistic process. Whatever happens, happens, and we have no control. But if the coin is much larger, the process becomes more predictable and deterministic. If we flip a very large coin at such-and-such a speed and angle, it will turn over 3 and 1/4 times before landing on heads. On the other hand, if the billiard ball becomes a very small ball bearing, its trajectory becomes more iffy and stochastic.

Something similar holds, I think, for the notion of "I" and biographies. When the perceived scale is small enough, the "I" seems more smeared and probabilistic and we feel more like a small coin or a tiny billiard ball. Chronicling the tiny minutiae of a life risks overfitting it and making it appear more iffy and stochastic than does a view of its broader outlines. If viewed from a distance with only gross aspects apparent, our "I" is seen as a source of more statistically predictable intention and appears to be a very large coin or billiard ball. Contrariwise, people often seem to themselves to be more conflicted than they do to other people. Autobiographies, in particular, display this indecisiveness and conflict. Again from a sufficiently removed position, however, almost everything can be seen as unsurprising.

Germane is the following phenomenon: In predicting what a person will decide on a personal issue, it is often very important to keep this predictive "information" away from the decider, else it change her decision. The quotation marks around "information" are meant to indicate that this peculiar type of information loses its value, becomes old news, if it is given to the person whose decisions are being predicted. The information, while it may be correct and true, is not universal. The onlooker and the deciding agent have complementary and irreconcilable viewpoints. The agent is unsure, a small "I" buffeted by a complex world. To the onlooker, however, the agent is a big "I" whose actions are often more foreseeable.

Whatever our level of analysis, predicting people and hence writing (midlife) biographies of them is quite difficult, made more so by our tendency to pontificate about large issues like hedgehogs rather than note the details like foxes. Research by political scientist Philip Tetlock and others shows that foxes are better at predicting than hedgehogs, who try to fit everything into the same Procrustean bed.[14] Apologies to Archilochus via Isaiah Berlin: "The fox knows many things, but the hedgehog knows one big thing."[15]

More generally, how things change with size and scale is better left to the fox than the hedgehog. The answer usually depends on many different factors, but generally one should always buy the big meatballs.

Chapter 10

MY STOCK LOSS, HYPOCRISY, AND A CARD TRICK

MY STOCK LOSS AND A FEW PITFALLS OF NARRATIVE LOGIC

Predictability and the stock market are words that don't always play nicely together. My experience with WorldCom stock in 2000–2001 is one of a gazillion others that bear this out. I'd always been and am now once again a firm believer in investing in broadly based index funds precisely because there is no convincing evidence that anyone can consistently predict the market, which is not to say that someone (brokers, for example, who do well whatever the market does) can't make a lot of money from it. The stocks and funds that do well one year (or the athletes featured on the cover of *Sports Illustrated*) usually fare much less well the next year. This is an instance of regression to the mean whereby an extreme measurement of a quantity dependent on many factors is usually followed by measurements closer to the average. Given this natural regression, highly paid fund managers and stock-pickers who charge hefty fees are, whether intentionally or not, conducting a sort of con game. Be that as it may, I acted in a way quite contrary to these beliefs and took temporary leave of my senses at the time and invested gradually, but heavily and foolishly, in WorldCom. I even borrowed from a line of credit on our house despite being aware that fraud and hyperbole were especially prevalent in the market at that time. What I didn't know was that some analysts were lying to prop up the price and reputation of WorldCom, whose CEO, Bernie Ebbers, went to jail for his efforts.[1]

The bottom line is that I lost my shirt (and one of my socks too) and afterward even wrote about my intense but virtual relationship with Ebbers for the *Wall Street Journal*. A natural reaction to the vagaries of chance is an attempt at control, so, Herzog-like, I wrote Ebbers and others a quixotic e-mail suggesting they promote the company more effectively.[2] I clearly had succumbed to confirmation bias, looking disproportionately eagerly for what made my investment look good and essentially ignoring everything that made it look bad. This wasn't difficult, given the stellar reports and strong buy recommendations that WorldCom's rather sketchy analysts kept bringing forth.[3]

Let me digress for the next two paragraphs about the folly of believing someone—politician, biographer, stock analyst—who is possibly lying when he or she is supported by other possible liars. If such a person is dishonest about some issue most of the time, then having his or her statements supported by someone else who is also dishonest most of the time actually decreases the likelihood they are telling the truth. This isn't surprising, perhaps, but let's look at the bare bones of such a situation involving just two people, Alice and Bob. Each of them tells the truth independently and at random 1/4th of the time, lying 3/4th of the time. Say Alice makes a statement. The probability that it is true is, by assumption, 1/4. Then Bob may "support" her, but not intentionally. He, like Alice, is just a psychopath who enjoys lying but randomly says something true 1/4th of the time, and he happens to say that Alice's statement is true. Now, given that Bob supports it, what is the conditional probability that Alice's statement is true?

First we ask how probable it is that Alice utters a true statement and Bob makes a true statement of support. Since they both independently tell the truth 1/4 of the time, these events will both turn out to be true 1/16th of the time (1/4 × 1/4). Now we ask how probable it is that Bob will make a statement of support. Since

Bob will be supportive when either both he and Alice tell the truth or when they both lie, the probability of this is 10/16 [1/4 × 1/4 (if both telling truth) + 3/4 × 3/4 (if both lying)]. Thus the probability that Alice is telling the truth given that Bob supports her is 1/10 (the ratio of 1/16 to 10/16). This scenario can be made more realistic, but the moral is clear: Confirmation of a very dishonest person's unreliable statement by another very dishonest person makes the statement even less reliable.

To get back to WorldCom, the mutually supportive crew around the company was very dishonest. Had WorldCom performed the way it was "supposed to" behave, especially after its price had declined and it appeared to be greatly undervalued, I would have made a lot of money and been seen as perspicacious and perhaps even gutsy. Instead I was … well, I'll skip the self-flagellation. Happily, I went on to write a successful book, *A Mathematician Plays the Stock Market*, which used stories and vignettes to explain and vivify various mathematical aspects of the market.[4] It resulted in my earning back much of the money I had lost, but, though money is fungible, equals can't necessarily be substituted for equals. Did I take a calculated gamble and then make lemonade out of lemons, or did I simply do something stupid and greedy? It's my life and even I can't answer this. Why should I put much faith in biographers' evaluations of their subjects' situations? And do I have advice to young people about life or the market? Of course not.

This debacle suggests a few general philosophical considerations in the writing of biographies and narratives of all sorts. One relevant notion is what are called intensional (with an *s*) contexts, involving verbs like *think*, *believe*, *feel*, and (as the above little vignette suggests) *wager*. In such contexts we can't always substitute equals for equals. By contrast, we can always do so in what are called extensional contexts such as mathematics. This is a fairly deep difference between narrative logic and mathematical logic. In mathematics, the substituting of equals for equals does not change the truth of statements. That is, whether we refer to 4, the cube root of 64, the smallest integer bigger than pi, or to 2 squared in mathematical contexts, doesn't affect the truth of our theorem or the validity of our calculation.

This is not the case in narrative contexts such as biography. In countless stories, both fictional and true, we know a fact F about X but don't know that the fact F characterizes Y as well because we don't know that X is the same person (or thing or stock) as Y. For example, Lois Lane knows that Superman can fly, but she doesn't know that Clark Kent can fly even though Superman equals Clark Kent.

Mistaken identity and simple ignorance are only small illustrations of this phenomenon, however. Someone may believe (note the verb), for a slightly different example, that Copenhagen is in Norway, but that person would not believe that the capital of Denmark is in Norway simply because Copenhagen equals the capital of Denmark. Substitution of equals for equals cannot be made. (An apocryphal story has it that Ronald Reagan believed Copenhagen was in Norway.[5]) The distinction between intensional and extensional is not unrelated to that between an action (say, pointing) and a movement (say, twitching or raising one's arm).

Since history and biographies deal largely with intensional contexts, the leverage of facts is somewhat limited. Consider any personal or historical event and substitute for incidents and entities in it any extensionally equivalent ones that come to mind. The result will likely be humorous or absurd, like substituting John Quincy Adams's inauguration day for any reference to one's wedding day. One would sound pretty silly proclaiming, "Ah, one of the happiest days of my life was the 137th anniversary of John Quincy Adams's inauguration day" even if the latter were indeed your wedding day. Or "I was so proud of my son on the second anniversary of 9/11." Is the boy a terrorist sympathizer or did he get accepted to medical school on that date?

More generally, our view of the event and the verdict of history on it depends to an extent on which extensionally equivalent characterization we choose. And which characterization we choose depends on many things, including our particular psychologies, the general historical context, and the history subsequent to the event in question.

Some general epistemological problems associated with narratives in general yield interesting special cases when phrased in terms of stock markets and biographies. Biographers (investors) claim to know many things about their subjects' lives (their stocks' worth). Sometimes their claims are true, sometimes false, but even when true, the claims don't necessarily constitute real knowledge. They could be mere speculation that happened to be true or, more prosaically, a misreading of the evidence. More interestingly, the claims might run afoul of a paradox due to Edmund Gettier, who showed that the ancient tripartite definition of knowledge as (i) justified, (ii) true, (iii) belief is not enough to ensure real knowledge.[6]

Imagine, for example, that a biographer reports that his subject X and another person Y were the only applicants for a certain position at a prestigious institute of mathematics. Suppose further that the biographer noted that though Y had a very serious, possibly disabling medical condition, evidence strongly suggested that Y was hired. The evidence might be that the institute badly needed someone with Y's specialty, but not X's, and that the director obviously liked Y. The biographer also noted that X was Jewish, the director was an anti-Semite, and the director and his secretary cackled maliciously after X was interviewed. Unknown to the biographer writing about X and the incident years later, however, is the fact that X, being the stronger candidate, was offered the position after all but decided to quietly turn it down. It was then offered to Y, who accepted it. The biographer asks with understandable indignation: why was the person with the very serious medical condition offered the job? But also unknown to the biographer is that X suffered from the same medical condition. Thus the biographer's claim that the person offered the job suffered from the serious medical condition is certainly justified, it is true, and the biographer believes it. But does the biographer know it? Not really. As in this case, justified true belief does not always equate to knowledge.

An autobiographical example prompted the above discussion. Near an elementary school one Saturday when I was about thirteen I was hitting a baseball to a friend of mine. A couple of times I really caught hold of the ball, and it bounced off the brick wall of the building. I laughed as did a couple of the school's students who were watching. Afterward my friend and I walked to my house and again crossed paths with the two young kids. Later that afternoon my brother Paul was throwing a ball against the same wall of the school building with his friends and shattered a window. The broken window was noted Monday morning, and the principal asked over the PA system if anybody knew how it happened. Of course, the eager young snitches told him about me hitting long fly balls that bounced off the building's wall and said they knew where I lived. The principal naturally believed that a kid who lived in my house broke the window. His belief was justified and it was true, but he certainly didn't know it.

It's not hard to note or even invent titillating biographical tidbits that might run afoul of Gettier's paradox or problems with intensional contexts or garden-variety mistaken identity (justified beliefs that, unlike Gettier situations, are false). A couple of examples: She's in the Catskills at a Ukrainian resort with her aunt's Botox doctor, the effect of whose injections on her would certainly rouse her aunt's suspicions. Or, he feigned a chronic kidney disorder as a ruse to regularly visit his paramour, a nephrologist, but when she tired of him, she called his wife to suggest dialysis. Clearly a similar story could be told about an investor and two possible investments X and Y.

For reasons such as those suggested above, in writing biographies it's difficult to compare a person's (or two people's) accomplishments or failings directly. Much depends on the beliefs, attitudes, and thoughts of the people involved, as well as on those of many others. Consider the following thought experiment: two fairly nondescript fellows have superficially similar lives until each undertakes some significant endeavor. Each endeavor is worthy and as likely to succeed as the other, but one endeavor ultimately leads to good things for X and his family, friends, and society, but the other leads to bad things for Y and his family, friends, and society. It would seem that X and Y should receive comparable evaluations for their decision, but generally they won't. Unwarranted though it may be, X will be judged kindly, and Y harshly. Even events occurring long after the deaths of X and Y can affect their reputations. If Y's son turns out to be a serial murderer and X's turns out to discover a cure for cancer, the posthumous regard accorded X and Y will be impacted.

Unfortunately with regard to WorldCom I played the role of bad Y rather than good X. I didn't receive or deserve a jail term, but I did receive and deserve a large fine.

ONE CHEER FOR HYPOCRISY

The above rendition of the chapter of my life dealing with the WorldCom debacle may strike some as an instance of hypocrisy. Whether it is or it isn't, hypocrisy is a notion for which I'd like to offer a partial defense. One step above black-and-white, either-or thinking is one-dimensional thinking, but it's a very small step. I'm reminded of various companies' TV advertisements that suggest all properties of a bed can be determined by its "sleep number." This, of course, is fatuous, but not as much as assuming that every political position can be arrayed along a liberal-conservative spectrum. Once one accepts this one-dimensional thinking, one naturally looks for apostasies and hypocrisies, but certainly not for real understanding. As I've noted, I've received my share of e-mails from people who've written (or screeched) that I'm a hypocrite because of some article, book, or column of mine, say, one recommending a cost-benefit analysis of something they, and they thought I, held sacrosanct. For some reason, these e-mails often begin ominously with "Dear Sir" or "My Dear Sir," which is almost as bad as "I'm praying for you."

Conventional understandings would suggest that I hold liberal positions on most issues, but I've known many "liberals" (myself included) as well as "conservatives" whose private actions and beliefs on some issues were on the opposite end of a spectrum (assuming that sometimes there is such a thing as a spectrum) from their public ones. As such they (and I) are often judged to be hypocritical. Examples might be environmentalists who don't recycle, libertines who rail against porn, gays opposed to marriage between transgendered people, gun-control advocates who have an arsenal of high-powered weapons in their basements, etcetera. Are these people hypocritical, as their biographers might be strongly tempted to say, or is it just easier to note their apparent conflicts than it is with other less "well-defined" people?

One common definition of *hypocrisy* is "the practice of professing beliefs, feelings, or virtues that one does not hold or possess."[7] (I wonder if one can be said to be hypocritical if one professes vices that one does not hold or possess.) It is usually considered to be a bad trait, and indeed it usually is, but it's rarely considered to sometimes be a necessary one. I think hypocrisy is occasionally unavoidable, and one of my reasons for thinking this is mathematical.

You may want to just skim this paragraph about the so-called *Entscheidungsproblem* in mathematical logic. It is the question of whether there exists an algorithm, that is, a well-defined recipe, for deciding whether or not a statement in the spare formal language of predicate logic is universally valid. Short answer: Nope, this isn't possible. In the '30s it was proved by Alonzo Church and Alan Turing that there is no algorithm such that, if given a statement, the algorithm will always respond yes if it is universally valid, or no if it is not.[8] The problem in a primitive form goes back to Gottfried Leibniz and his dream of a universal reasoning machine.[9]

Related limitations hold for the even sparer, more limited language of propositional logic. The problem here is whether, given some complicated combination of simple propositions P, Q, R, ... connected only by *and*, *or*, and *not*, there is any way at all to assign truth or falsity to the simple propositions in such a way that their complicated combination is true. The problem in this case is decidable, but it, the so-called Boolean satisfiability problem, is still NP-complete (suffice it to say, an extremely difficult logical problem).[10]

Without the mathematical jargon, the point is that it's not always easy or even possible to determine quickly if some collection of statements is ever satisfiable, or universally valid, or provable. We probably all subscribe to inconsistent or unsatisfiable collections of statements and hence are at times hypocritical—sometimes knowingly so, sometimes not. This is all the more the case when we go beyond the formal languages and rules of mathematical logic to natural languages that admit of connotation, vagueness, and intensionality.

Sexual morality is one area particularly prone to hypocrisy. Inundated by sexual images, Viagra ads, and ubiquitous porn as well as by moralistic preaching and preening about fidelity, marriage, and "cheating," we might be excused—at least sometimes—if we're not sure how to avoid hypocrisy and achieve consistency. A similar example is the sometimes futile attempt to reconcile libertarian beliefs, attitudes toward prostitution and feminism, (ir)religious convictions, and economic or political ideologies. If you dislike someone, you can probably find a rationale for a charge of hypocrisy; if you like the same person, you can probably find a way to describe his or her positions as nuanced and thoughtful.

I've mentioned major issues here, but balancing one's behaviors and beliefs toward the people we're closest to is probably even more difficult since there are many more variables, incommensurable convictions, and competing desiderata involved in our personal biography. The bottom line is that for mathematical reasons as well as, of course, for social and psychological ones we ought to be more tolerant of what is, or appears to be, hypocrisy. More specifically, we don't need too many overly earnest biographers scolding their subjects about their foibles, especially when some of them are mathematically inevitable. Paraphrasing noted writer François de La Rochefoucauld, I think that hypocrisy is sometimes the homage that truthiness pays to truth.

KRUSKAL'S CARD TRICK AND COMMON DENOUEMENTS

Although I don't have a Facebook account, much preferring Twitter, I hear of and from people I knew a long while ago. More often than not I find that they, despite having chosen very different paths in life, have developed along those different paths in somewhat comparable ways. Given the huge disparities in their past lives, they nevertheless tend to react more and more alike with regard to participation in sports, the latest gadgets, enrolling in a course, or travel as they age, *age* being the crucial word.

More generally, most people's lives, no matter how significantly different they are and how much they've diverged since adolescence and how unpredictable their later beliefs, values, and achievements turn out to be, nevertheless often seem to acquire a similar shape as they age. In the language of abstract algebra, they evince a certain degree of symmetry and invariance. The latter are complementary notions. Something is symmetric to the extent that it is invariant under (or unchanged by) some sort of transformation.

Circles, for example, are invariant if rotated or reflected about a diameter. Ellipses are not invariant under these transformations, but are under some. Even if an ellipse is squashed, its center still bisects each of its diameters whatever their length. A topological figure is invariant under continuous stretching and bending, which is why coffee cups and donuts are topologically famously the same. The German mathematician Felix Klein suggested that theorems about geometrical figures might be classified according to whether or not they remain true when the figures are subjected to various changes and transformations. That is, given any specified collection of transformations (rigid motions in the plane, uniform compressions, projections), Klein asked what properties of figures remain invariant under these transformations.[11]

So what properties of people remain invariant despite the transformations brought about by very different life stories? As a start, note that people might seem to have a limitless number of choices early in their lives and, of course, they make different choices among them. Memories from this period are especially numerous and salient and constitute the above-mentioned reminiscence bump. Middle age brings a somewhat smaller set of options—marriage, children, a profession, office work, farming, factory job, business, and so on—which leads to growth along these occupational and familial lines, more specialized paths, and sparser memories as well as a narrowing of available options. Near the end of people's natural lives, it often seems that despite traversing quite different routes, there is a hard-to-define retrospective invariance to their outlooks.

To the extent this observation is true but not trivially so and can be made more precise and testable, historical factors, social conventions, psychological affinities, and biological and neurological factors help explain it. So, too, might the insights of so-called structural anthropology, which maintains that human characteristics are the same in every culture. There is, however, a possible mathematical reason for the phenomenon as well. It involves a mathematical card trick, one of my favorites, that illustrates how people might gradually fall into a similar sequence of development despite early quite obvious differences among them. People's sometimes lockstep development is, if not taken literally, a nice illustration of the trick and of the way mathstuff permeates our lives.

As I've written elsewhere, the card trick was invented by the physicist Martin Kruskal and can be most easily explained in terms of a deck of cards with the face cards removed.[12] Imagine two players, Dupe and Con Man. Con Man asks Dupe to pick a secret number between 1 and 10—say he picks X—and goes on to instruct Dupe to watch for the Xth card as Con Man one by one slowly turns over the cards in a well-shuffled deck. When the Xth card is reached—say it's a Y—Con Man explains that it becomes Dupe's new secret number, and Dupe is asked to watch for the Yth card after it, as Con Man continues to turn the cards over slowly one by one. When the Yth succeeding card turns up, its value—say it's Z—becomes Dupe's new secret number, and he is asked to continue in this way, watching for the Zth card after it for his new secret number, and so on.

For example, if Dupe first picks 7 as his secret number, he would watch for the 7th card in the deck as Con Man slowly turns the cards over. If the 7th card is a 5, his new secret number would become a 5, and he would watch for the 5th card after it. If the 5th card after this is an 8, then 8 would become his new secret number, and he would watch for the 8th card after it to determine his new secret number. As they near the end of the deck, Con Man turns over a card and announces, "This is your present secret number," and he is almost always correct.

The trick is purely mathematical. The deck is not marked or ordered, and there are no confederates, no sleight of hand, and no careful observation of Dupe's reactions as he watches the cards being turned over. So how does Con Man accomplish this feat? The answer is clever. At the beginning of the trick, Con Man picks his own secret number. He then silently follows the same instructions he's given to Dupe. If he picked a 3 as his secret number, he watches for the 3rd card and notes its value—say it's a 9—which becomes his new secret number. He then looks for the 9th card after it—say it's a 4—and that becomes his new secret number, and so on.

Even though there is only one chance in ten that Con Man's original secret number is the same as Dupe's original secret number, it is reasonable to assume and can be proved that sooner or later their secret numbers will coincide at some point during the process. (Two decks work better since the longer the sequence of cards, the more likely the secret numbers are to be the same at the end.) In other words, if two more or less random sequences of secret numbers between 1 and 10 are selected, sooner or later they will, simply by chance, lead to the same card. Furthermore, from that point on, the secret numbers will be identical since both Dupe and Con Man are using the same rule to generate new secret numbers from old. Thus all Con Man does is wait until he nears the end of the deck and then turns over the card corresponding to his last secret number, confident that by that point it will very probably be Dupe's secret number as well.

If you feel like checking this out on a table of random digits between 0 and 9, pick any digit, your first secret number, from the first line of the number array below. Find your second secret number from the first one and the rule above and continue in this way. Your last secret number will almost certainly be the last 6 in the fourth line of the array. Try it. Really, do.

11164363187516137674263227514315431264181922891792
21215917917683158678877543168793285436851973298468
11438444826655837649388829487612462418157189682977
36792262363326666583618819739523461367422285254564

Note that the trick works just as well if there is more than one Dupe or even if there is no Con Man at all (as long as the cards are turned over one by one by someone). If there are a large number of people and each picks his or her own initial secret number and generates a new one from the old one in accordance with the procedure above, it's quite likely that all of them will eventually have the same secret number and thereafter will move in lockstep until the cards run out.

If we complicate things and allow people's secret numbers to be determined in more convoluted ways from several of its preceding secret numbers instead of just from its immediate predecessor, and if we change the scenario from turning cards over one by one to some other sequential process, we see the potential for lockstep behavior on a large scale to develop naturally. If, for a different sort of example, many investors use the same computer software, i.e., follow the same rules for high-frequency trading, it is quite conceivable that some variant of the above might result whatever the investors' initial positions.

In *Once Upon a Number* I even proposed a religious hoax inspired by the Kruskal card trick and a desire to mock the fatuous so-called Bible code. Instead of using numbers directly I suggested using the number of letters in successive words in a passage from the Bible. The passage would be such that the Kruskal procedure always resulted in the last word being in one way or another particularly evocative to a superstitious person.[13]

To illustrate, I'll use the same variant of the trick I used for a recent Math Awareness Month, but this time applied to the Declaration of Independence. Pick *any* word from the first two lines of the Declaration of Independence (see the excerpt below). Call it your special word. Count the number of letters in it and move forward that number of words to your next special word. For example, if you picked the word *course*, which has six letters, as your special word, then you would move ahead six words to *necessary*, which becomes your next special word. It has nine letters, so you would move ahead nine words, bringing you to your next special word, *which*. Continue to follow this rule until you can't go any further. No matter what word you pick initially, your last special word will be *happiness*.

Here's the example.

> When in the *Course* of human events, it becomes *necessary* for one people to dissolve the political bands *which* have connected them with *another*, and to assume among the powers *of* the *earth*, the separate and equal *station* to which the Laws of Nature *and* of Nature's *God* entitle them, *a decent* respect to the opinions of *mankind* requires that they should declare the *causes* which impel them to the *separation*. We hold these truths to be self-evident, that all *men* are created *equal*, that they are endowed *by* their *Creator* with certain unalienable Rights, that among *these* are Life, Liberty and *the* pursuit of *Happiness*.

You might want to try this starting with an early word in the passage different from *course* and check to see that you still end up with *Happiness*. So does the Declaration guarantee happiness?

In biographies, the role of the cards might be played by the randomly presented age-appropriate possibilities life offers or by significant current events or by other shorter-term recurring outcomes, and the more complicated rules followed by the subjects are their natural reactions to these possibilities. As mentioned, social conventions and psychological affinities underlie these reactions. There are obviously many more than ten of them at each decision point in subjects' lives, and their reactions are not nearly as determined by their previous choices, although, as people age, their reactions do grow less plastic and more determined. Reflecting this difference, the number of possibilities decreases as time goes on. This is comparable to a variant of the card trick in which a couple of cards, say, all the 3s and 6s are removed from the second half of the deck and all the 4s and 9s are removed from the last quarter of the deck.

The trick and more realistic variants can be modeled by the notion from probability theory of a so-called absorbing Markov chain.[14] It is an interesting mathematical problem to discover the most general conditions leading to lockstep behavior. Alas, the last card for all of us is that we're absorbed by the earth.

Before that, however, let's examine some other general aspects of biography.

Chapter 11

BIOGRAPHIES: VERSTEHEN OR SUPERFICIAL

CONSCIOUSNESS, BIOGRAPHIES, AND SHMATA

As I've expressed frequently herein, for a variety of reasons I find biographies to be somewhat inadequate in getting at what a person is like. I've cited paradoxes, counterfactual assertions, straightforward lies, mistaken assumptions, bizarre interpretations, statistical solecisms, cognitive foibles, and just plain wrongness, but much of what I write in this book can be stated rather succinctly. People in general and biographers in particular are often, to use a common slang expression, full of it (the "sh" is generally omitted in polite company). The phrase is common because we apply it so often to others, but those with any self-awareness at all must realize that the admittedly vague phrase often applies to themselves as well. It definitely applies to me at times. In rereading the autobiographical vignettes of this book, for example, I'm discomfited by the feeling that, though true, parts of them indicate that I'm at least 14.3 percent full of it.

Three quotes attributed to authors Mark Twain, Paul Auster, and Rebecca West, respectively, compellingly express similar sentiments.

> What a wee little part of a person's life are his acts and his words! His real life is led in his head, and is known to none but himself. ... Biographies are but the clothes and buttons of the man. The biography of the man himself cannot be written.[1]

Every life is inexplicable, I kept telling myself. No matter how many facts are told, no matter how many details are given, the essential thing resists telling. To say that so and so was born here and went there, that he did this and did that, that he married this woman and had these children, that he lived, that he died, that he left behind these books or this battle or that bridge—none of that tells us very much.[2]

Just how difficult it is to write biography can be reckoned by anybody who sits down and considers just how many people know the real truth about his or her love affairs.[3]

Of course, we shouldn't conclude that lives are completely inscrutable or that clothes and buttons aren't sometimes fascinating. It's just that biographies are usually heavy on movement, anecdotes, official acts, and superficial connections among the protagonist and other people, but light on insight into the subject's real attitudes and personal worldview.

This is especially true of the dynamic and chaotic relations (these adjectives are intended in both their everyday and their mathematical senses) among families, friends, and colleagues, the open secrets that people know while remaining ignorant of others' knowledge of same (the distinction mentioned earlier between mutual knowledge and common knowledge), the sturm und drang (storm and stress) of coming of age, and their associated endearing memories and enduring resentments. Knowing my limitations, I haven't even attempted to limn these topics here, where I've focused largely on myself and mathematics.

The quotes above clearly suggest this and seem to be pointing to the lack of Verstehen in most, if not all, biographies. *Verstehen* is a German word meaning "to understand," but in sociology it's often used in a special way to refer to a sort of understanding that requires interpretation and empathetic involvement. The under-standing is not completely objective but requires that you appre-ciate how the person in question sees the world, view the person as a subject and not merely as an object of study, even identify with his or her point of view. (The term also carries a good deal of unconvincing metaphysical baggage that is not relevant here.)

Verstehen is obviously important in biography since how biographers describe an event or person is often critical, and their ability to "see as" the other person does is essential (and at best very rare). Even simple actions, not to mention complex sequences of actions, illustrate this. Here is an example I've used before: If a man touches his hand to his forehead, we may see this simply as an indication that his temple is throbbing. We may also see the gesture as a signal from a football coach to the quarterback. Then again, we may infer that the man is trying to hide his guilt by appearing nonchalant, that it is simply a habit of his, that he is worried about getting dust in his eye, or indefinitely many other things depending on indefinitely many perspectives we might have and on the indefinitely many human contexts in which we might find ourselves. Describing the movement in only physical terms—how fast and at what angle his arm moves, the physiological correlates of the movement, etcetera—obviously does not provide Verstehen. Even young children understand this distinction since they might deny hitting someone and claim they were just swinging their hands when somebody's face walked into them. My son Daniel, now a very good lawyer, once made this claim after he'd hit my daughter Leah. She, ever savvy even as a toddler, wasn't at all taken in by this subterfuge.

People generally have enough Verstehen of their spouses (plural, spice?) to know that how he or she sighs can signal a change in mood or a change in attitude toward someone or something, especially when a verbal acknowledgment is too hard to make immediately. Sometimes my wife responds to the merest twitch of my eyebrow with something like "Why do you think that?" (Of course, she's often wrong about what I was thinking.) Likewise, when one knows a person intimately, that person's facial expressions, interactions, and the incidents they choose to relate can sometimes also hint at immensely complicated, almost incompressible, yet slightly ambiguous stories. This complex indeterminateness can even arise when observing or writing about oneself. Seeing an account of my actions or thoughts on the page sometimes makes me reconsider who this "I" is that I carry around all day and who seems to always be speaking for me. For the record I'm usually quite fond of the guy, but I wish he'd shut up occasionally, since he sometimes gets things a little wrong.

An account of a life that relied almost totally on Verstehen rather than exclusively on clothes and buttons would occupy its subject's consciousness, explore his or her relations with significant others, and would be a rather electric matrix of diversion and digression. Messy—like life (and this book). That is, details, both big (say, political or economic concerns) and small (say, annoyance at the inclusion of nutmeg in a dish or a momentarily all-consuming hangnail), on matters both critical and trivial would follow one another unpredictably through the chronicle.

The story and its multilayered details—like endlessly jagged coastlines, or creased and cracked mountain surfaces, or the whorls and eddies of turbulent water, or a host of other "fractured" phenomena—might suggest to you the mathematical metaphor of a fractal in some sort of "cognitive space." After all, characteristic of fractals is a kind of relative simplicity that is nevertheless complex-seeming, unlimited branching and diverging, and recursive self-similarity, whereby individual instances are defined in terms of their predecessors and all have a somewhat similar look or feel no matter on what scale one views them. (The latter property explains the following joke about the founder of fractal geometry. What does the B in Benoit B. Mandelbrot stand for? Answer: It stands for Benoit B. Mandelbrot, the "father of fractals," the sequence of imbedded names constituting a kind of fractal. A Russian doll, another example, is one that is full of ever-smaller Russian dolls.)

By exhibiting—that is, showing as well as telling—some of the inexhaustibility of such self-reflective rambling (some of it truly complex in a sense I'll get to shortly and not at all fractal-like), such a biography would demonstrate the flavor of a particular fractal human consciousness. Many of its little episodes and vignettes would resemble nothing more than teenage intrigues ("I didn't tell Mimi because she would certainly tell Helen, who might tell Dick, and I didn't want him to hear something other than what Ellen and Jerry had told him before they knew the extent of it.") The protagonist's thoughts would be dense with life and would give the reader a substantial degree of Verstehen for the person. Although they're not biographies, some close literary cousins of this approach might be James Joyce's *Ulysses* or Marcel Proust's *Remembrance of Things Past*. In fact novels, though fictional, are often better at imparting Verstehen than are biographies.

But even if it made sense and it was possible to really grok a person, to "get inside the head of" the subject and know and think the way he or she did, we would still have two huge problems. Since Freud, countless psychological and neurological studies have demonstrated repeatedly that people, biographical subjects or not, are not aware of many of their own motivations. They (we) do things that they (we) don't understand. Clothes and buttons certainly aren't sufficient in fashioning a biography, although they are often interesting.

Our common folk psychology, which is indispensable in everyday life, and even Verstehen fall short if the subject is unaware of why he or she did something. (Once again, "did" may be unjustified; people see agents everywhere even when the movements are not at all intentional.) The tired metaphor of the mind being an iceberg most of which is under the surface is misleading in that it suggests that the conscious and unconscious parts are somehow connected, somehow continuous.[4] Not so. This unconscious unawareness of many of our actions and motivations would seem to limit the value of both commonsense psychology and Verstehen in understanding a person's state of mind.

Secondly, whether of the clothes and buttons or even of the much more full-bodied and vibrant Verstehen variety, biographies fall short for another quite different reason: we're just too complex to be captured fully. The informal notion of complexity can sometimes be modeled by the formal mathematical notion I mentioned earlier, which is due to both the computer scientist Gregory Chaitin in his book *The Limits of Mathematics* and to the Russian mathematician Andrey Kolmogorov.

Please indulge me as I digress to establish a few details of the notion. The definition is simple and elegant. The complexity of a sequence of bits (0s or 1s) is roughly the length of the shortest program needed to generate the sequence in question. The repetitive sequence 011011011011 ... 11011011 ... is, for example, not very complex since a very short program generates it: Print 011 and repeat however many times are needed. More complex sequences require longer programs to generate, and a sequence of bits is said to be random if the shortest program generating it is incompressible and essentially as long as the sequence itself, say, a sequence like 01001000010110100111100101000001011. ... A program generating such a sequence can do little more than repeat it: Print 01001000010110100111001010. ... It simply spits out the sequence itself since it can't compress it. If I can resort to base reductionism for a second, I note, given that everything—DNA, music, us—can be encoded in such sequences, these definitions apply to us as well.

I mentioned music for a reason. The computer scientist Donald Knuth illustrates this notion by examining common ways to reduce the complexity of songs in order to appeal to children or to ease the burden on memory. Having a refrain that occurs throughout a song is one way of doing so. Consider "Old MacDonald Had a Farm." The familiar "and on that farm he had an (animal name), E-I-E-I-O, with an (animal noise twice) here and an (animal noise twice) there, here an (animal noise), there an (animal noise), everywhere an (animal noise twice)" followed by "Old MacDonald had a farm, E-I-E-I-O" leads to a considerable reduction in complexity.[5] A way to reduce complexity even further is exemplified by songs such as

"99 Bottles of Beer on the Wall," where the song proceeds downward through 98, 97, . . . 3, 2, 1. Finally, consider "That's the way I like it, uh-huh, uh-huh" repeated over and over again.[6]

Chaitin developed the notion of complexity in part to give an alternative proof of Kurt Gödel's famous incompleteness theorems. Gist of the proof: We can't expect (the complexity) of 5 pounds of axioms to yield (the complexity of) 10 pounds of theorems. In order to generate a certain degree of complexity—10 "pounds" worth, say—what is needed is more than 10 pounds worth of complexity.

This is also why explaining the complexity of the world by invoking the even greater complexity of a deity explains nothing; it replaces one imponderable with an even bigger one. It also suggests that to the extent we ourselves are formal systems, we are subject to the same existential incompleteness.

Not unrelated is the Berry paradox, first published in 1908 by Bertrand Russell. This paradoxical twenty-word sentence asks us to consider the following task: "Find the smallest whole number that requires, in order to be specified, more words than there are in this sentence."[7] Examples such as "number of hairs on my head," "number of different states of a Rubik's Cube" and "speed of light in millimeters per century" each specify, using fewer than twenty words, some particular whole number. The paradoxical nature of the task becomes clear when we realize that the Berry paradox specifies a particular whole number that, by its very definition, the sentence contains too few words to specify.

End of digression. For my purposes here I note that lying between the extremes of compressibility mentioned above is where life resides, where some compression is possible—via fractals, for example, or in the case of songs via refrains—and where some parts are relatively incompressible. This is a big interval. Some people's lives are simpler than others, and their descriptions admit a higher degree of summarization and compression, something captured by philosopher Derek Parfit's description of such lives as containing "only Muzak and potatoes";[8] a description of other more complicated lives would of necessity be much longer. Of course, nobody (not even Tristram Shandy) could write or read a full biography that left out no significant facts and events. The bottom line is that the complexity of human lives is too great to be captured in any reasonably full way, and a serviceable approximation of Verstehen is the best we can hope for. The map can never be as complete as the territory.

These considerations and limitations are more salient for authors attempting biographies of people from very different cultural backgrounds (anthropologists also use the term Verstehen) or historical periods (explaining the mythical Achilles' motivation in the *Iliad*, for example). Going much, much further afield to "biographies" of other species suggests a more theoretical matter and brings to mind the philosopher Thomas Nagel's famous article "What Is It Like to Be a Bat?"[9]

In his somewhat polemical article Nagel argues that consciousness has an irreducibly subjective aspect and states that "an organism has conscious mental states if and only if there is something that it is like to be that organism—something it is like for the organism."[10] Unlike the situation with physical concepts, to even have the notion of a mental concept requires, according to Nagel, that we be directly acquainted with it. Mental states are not best understood from an objective perch. For such states there is no such thing; the person is the perch. Being a person entails having a subjective perch from which to view the world, Nagel argues, and any attempt to objectively account for or explain away this perch, this personal perspective, will of necessity omit what we would like to understand.[11]

More controversially Nagel adds that a real grasp of consciousness and mental concepts might require a profound change in our understanding of both the mental and the physical.[12] Although they appear to be essentially different, that may be only because of our present state of knowledge. To write a "real" biography, to allow others to understand what it's like, beyond clothes and buttons, to actually be an Abraham Lincoln, a Thomas Edison, or an Aung San Suu Kyi might require an un-anticipatable advance in our understanding of physics and consciousness.

Although no compelling physical evidence exists that consciousness is anything more than a purely neurological phenomenon, I have to grant that it may be. Entering foolishly into nearly (or totally) nonsensical speculation inspired by Nagel, I note that consciousness, the ghost in the physical machine of our brains, might, like the Higgs boson, be almost impossible to detect and yet be a pervasive field of energy. Like the Higgs field, the "consciousness field" might impart consciousness to the right sorts of entities—sufficiently complex, appropriately structured, suitably self-referential physical objects such as human and animal brains.

And these considerations and speculations prompted by Nagel, physicist Freeman Dyson, and others finally bring me to Shmata. One day my wife brought home a small, shaggy dog, a Bichon Frise, from a kill shelter. Someone had for some hard-to-fathom reason taken her there. The dog was white and fluffy but very dirty and matted, so my wife named her Shmata, Yiddish for *rag*. Gradually Shmata overcame her abusive background and grew into the affectionate, smart, and neurotic dog we now know. She's full of idiosyncrasies and relates to my wife and me in two entirely different ways. She licks me constantly but wants my wife to always be petting her. She barks loudly and jumps up and down like an insane jack-in-the-box if we get out of the car and we're a nanosecond late in letting her out. She knows dozens of words and phrases, and she is omnivorous—and deceptive about it too when she discovers something on the street she knows we'd object to her eating. She also loves matzoh, so the long, formal version of her name is Shmata the Matzoh-Eating Canine-ite. She is as deserving of a biographical sketch (even if it's only clothes and buttons) as any other sentient being. Hence this paragraph that I'm sure falls far short of answering the question: What Is It Like to Be Shmata?

LEAH AND DANIEL, MY GRANDSONS AND I

I do have a vastly better feel for what it's like to be my daughter Leah or my son Daniel than I do about Shmata, but I don't wish to write much about them. They didn't sign up for embarrassment by their writerly father and need to be the authors of their own life stories. I do want to say that they are mensches (or, ignoring etymology, a mensch and a womensch). They're sensitive, kind, smart, and level-headed and are an unalloyed joy (or maybe only exceedingly rarely alloyed).

Since this book is mathematically tinged, I will say something about their relation to the subject. When they were very young, I taught them to answer the questions "What is the derivative of x squared and, secondly, what is the derivative of e to the x?" They would respond in their toddler voices, "2x is the derivative of x squared and e to the x is the derivative of e to the x" and then would, like me, be amused by the momentary surprise their answers elicited from my mathematical colleagues. Leah and Daniel knew it was a joke and, unlike more than a few of my calculus students, were under no illusion that they understood anything. More generally, they liked the mathematical puzzles—both standard (pico, fermi, zilch) and idiosyncratic—that I would pose to them while driving from Philadelphia to their grandparents' apartment in New York on the long, boring New Jersey Turnpike. They also both excelled at mathematics in school, my son receiving a master's degree in it before opting for law school over a PhD in the subject, and my daughter majoring in engineering in college before switching to English and related endeavors.

I'll limit myself to their math prowess and skip the accounts of family trips in this country as well as to England, France, Greece, Israel, Thailand, Japan, the jokes and code words that living together engender, playing ball against Sandy Run Middle School, the countless conversations—most silly, some serious—and simply repeat that they're great and probably will be embarrassed to read even these few sentences.

Children often have children, and so a few words about my grandson Theo (he of the earlier Cheerios mention), who is too young to complain about my writing of his skill as a reverse Roomba. Like most toddlers he is a fount of either wisdom or unwisdom. I'm not sure which. When he finally started to use the potty, he told Leah that a lot of it got on the floor but that he was "okay with that," a phrase he had apparently heard somewhere. Some of his little friends in the play group were discussing their pets and asked if he had a pet. He replied earnestly, "No, just Charlie, my brother." Charlie, who calls himself a very mean pirate (except when he has a cold and then calls himself a sick pirate) despite being ever agreeable, would probably have said "yah" had he heard the exchange.

One viscerally loves one's children and grandchildren for their own sakes (although the feeling necessarily becomes abstract with, say, one's great-great-great-great-great grandchildren). Grandchildren, children, their spouses (Andy and Marie, in my lucky case), other family members, friends, colleagues, and fellow citizens allow us a way to extend the short thread that is our life by weaving it into a larger patch that is in turn part of the whole social fabric. It's not just about us. That's the idea anyway, and it's compelling and appealing, but we might look at the metaphor in a more jaundiced way: individually we're just pieces of lint. Not sure if that's a piece of wisdom or unwisdom. Theo, Charlie, and I are just trying to make sense of things as best we can at our respective stages in life. As I write this their cousin Max, recently arrived on the scene, has just begun his long and complex trajectory.

The thought of eventually fading away and being survived by one's own and the larger social family suggests to me what will happen when the earth's low-lying islands become submerged because of global warming and the melting of the ice caps. This prospect is fairly imminent, as the Maldives and many others surely won't make it for many more decades. The prudent leaders from some of these islands are now buying land on other nearby islands or on the mainland to relocate their people there when their home island goes under. They are also arguing to maintain all the fishing and natural resource rights not only on the home island itself but also in a twelve-mile buffer zone around the soon-to-be-nonexistent island. The less than watertight analogy is to the transference of one's estate or legacy to one's children as we, like these islands, get washed away in the sands of time.

To counter the latter cliché, I'll end with this wise quote from the poet Robert Frost: "In three words I can sum up everything I've learned about life: it goes on."[13]

GOMPERTZ'S LAW OF HUMAN MORTALITY AND LIFE SPAN

Becoming a grandparent or simply getting older usually brings about a keener sense of mortality. Few unasked questions are more human than: How much longer do I have? How many more times will I travel here, eat there, do this or that thing I've enjoyed (or simply endured) doing?

Excluding external causes of death, we can give a general answer that is somewhat surprising. Whatever the probability p that you, an adult, will die during the next year, 8 years from now the probability you'll die during the following year is 2p or twice as great. For example, if you're 40 years old and your chances of dying during the next year are about 1 in 800, when you reach 48, your chances of dying during the following year would increase to about 1 in 400, and when you reach 56, your chances of dying during the following year would rise to approximately 1 in 200.

The scope of this relation, which was discovered by Benjamin Gompertz, a nineteenth-century British actuary, extends across borders and time periods. Even if the life span in a given country during a given historical period is much lower, the probability of a person's dying during the next year still doubles every 8 years. (This phenomenon also brings to mind the Kruskal lockstep progressions mentioned earlier.)

Gompertz observed that the annual mortality rate rises exponentially, which implies that the probability of surviving to a given age shrinks rapidly. The phenomenon is not unrelated to the rule of 70 in financial matters. If your money is growing at a rate of r percent, it will double in 70/r years. It's not at all clear why this mortality relation should hold.

As physicist Brian Skinner has observed, Gompertz's law, though somewhat mysterious, suggests that our exponentially increasing mortality is the result of a "built-in expiration date." He further observed that various theories of the causes of death do not give rise to known mortality tables.[14]

If, for example, we were to subscribe to a constant mortality rate of 1 in 80, the latter number being the average life span in the United States, this would mean your chances of dying next year would be 1 in 80, the chances the year after that also 1 in 80, and so on. This would result in an average life span of 80, to be sure, but with many people dying in their teens and twenties and a few living to age 300 or more.

Even if one makes this approach more realistic, for example by saying you have a 1 in 16 chance of suffering some injury or disability or deterioration and accumulating 5 of them kills you doesn't give real-life mortality tables. These assumptions also give rise to an average age of 80 (16×5), but with many people living past 150. Similar "fixes" also fail to yield, as Gompertz's law does, known mortality tables.

One model described by Skinner that does yield realistic life-span distributions is called the "cops and criminals" theory of the immune system. It is schematic and suggestive in something like the way Hofstadter's theory of simms and simmballs is (see chap. 3). If a sufficient number of circulating cops are patrolling your body, they can wipe out any criminals they happen upon. However, if the criminals are allowed to loiter too long, they build up fortresses that the cops can't penetrate, and disease or cancer develops. As the number of circulating cops decreases for whatever reason, their patrols become less frequent and criminal fortresses more numerous.[15]

This colorful model dovetails nicely with science writer George Johnson's abstract perspective on the inevitability of cancer.[16] The ultimate cause he argues is simple entropy. Over time DNA deteriorates or is miscopied or otherwise degraded, events whose inevitability is a consequence of the second law of thermodynamics. Once again we behave like mathstuff.

"Cops and criminals" is, of course, quite suggestive of an immune system fighting off assault, a model made all the more plausible by the fact that cancer incidence also seems to double every eight years à la Gompertz. Unlike cancers of the young, the many cancers of the elderly have remained and will remain recalcitrant, making getting rid of "criminal mutations" ultimately impossible. The problem is that mutations, criminal or not, are, as Johnson observes, the engine of evolution and without them, no complex life would exist. Sometimes and somehow whether because of glitches, a reduced police force, or whatever, some mutations begin to grow into cancer. Things fall apart, including us.[17]

Finally, I should mention a heuristic device that gives a reasonable estimate for your remaining life expectancy. Based on figures from the US government, it states that you can expect to live another 72 years minus 80 percent of your present age if you are under 85. If you are over 85, you can expect to live another 22 years minus 20 percent of your present age. For example, if you're 60 years old, your remaining life expectancy is 24 years.

Once again, no matter how we slice it, we fall apart.

Chapter 12

TRIPS, MEMORIES, AND BECOMING JADED

TOPOLOGY, TRAVEL, AND A THAI TAXI DRIVER

I've always loved travel, whether real or virtual, and feel that it's crucial to understanding oneself, at least to the extent that is possible. In some of the *Where's Waldo?* children's books, the task is to pick out the character Waldo from the dozens and dozens of other figures in the books' large colorful illustrations. This is made easier because of Waldo's distinctive dress and coloring. Traveling in a foreign country sometimes makes one feel like an anomalous Waldo, one's distinctive dress and coloring being largely mental. It's a truism that one often feels most American (French, Russian, Israeli, Argentinian, whatever) when one is immersed in a foreign culture.

Sheila and I, temporarily Waldo and Wenda, certainly felt this way in India. The state of Rajasthan in particular was truly amazing, truly overwhelming, the very definition of extreme incongruity. The place was very religious (superstitious), warm, colorful, noisy, and dirty. It was as absurd a place as I'd ever been, and it made Thailand, to which I've traveled with my wife and daughter and where I've lectured several times, seem like Norway and Thais like logical positivists by comparison. Superficial observations perhaps, but India struck me as a very hierarchical society, castes still playing a determinative role, and composed of many subgroups, religions, and cultures seemingly from different centuries, ranging from the modern, hi-tech one and Bollywood to the thousands of primitive medieval hamlets.

As much as I've enjoyed traveling with my wife and children and, as mentioned above, treasure the trove of common memories it creates (such as the silly loutraki dance whose name and steps we made up or the bizarrely humorous English on signs in Japan), traveling alone also has its appeal, especially in a country like India or Thailand with different traditions and mores. It is simultaneously exhilarating and anomie-inducing. The social and familial supports that usually surround you, replaced only in a limited way by their instantiations within you, are largely absent. You're swimming in a different cultural sea, and you can either try to maintain your normal stroke and swallow water unpleasantly, or else you can just float and enjoy the strange currents. Most likely you'll do a bit of both, hence the exhilaration and anomie.

A small example: A few years ago I traveled to Thailand to visit Jim Rakocy, an old and good friend of mine and expert in aquaponics, who had retired there. He lived an hour and a half south of Bangkok, so I took a taxi there that happened to be driven by a sixty-four-year-old who spoke good English. Being approximately the same age, we talked a bit about our lives, and Jumbo (the name he reported to me) told me he worked in a US army PX during the Vietnam War. We reminisced about the times, and he put on a CD he'd made consisting of pop songs from that era—"Itsy Bitsy Teeny Weeny Yellow Polka Dot Bikini," "Bird Dog," some songs by Joan Baez, and assorted plaintive love songs. He said the music brought him back to those simpler times and all the good times he'd had. His reveries about his experiences with sex workers were oddly thoughtful, empathetic, and introspective. Hyperbole to be sure, but the neologism *proustitution* occurred to me.

Discombobulated by my jet lag and Jumbo's stream-of-consciousness musings, what came to my mind, aside from nostalgia, was the subject of topology and a few of its theorems. As I mentioned in *Beyond Numeracy*,[1] Woody Allen once parodied the subject, writing that "fake rubber ink blots were originally eleven feet in diameter and fooled nobody. Later, however, a Swiss physicist proved that an object of a particular size could be reduced in size simply by 'making it smaller,' a discovery that revolutionized the fake ink blot business." Contrary to Allen's remark, however, is the fact that many of topology's theorems are quite counterintuitive and therefore perhaps are appealing to a mind reeling from an eighteen-hour flight and a mild nausea.

Or maybe I was just trying to focus on something familiar to me and structure-providing, Whatever the reason, this area of mathematics came to mind as did the thought that all arcs (and lives, to reiterate, are in a sense, simply arcs or trajectories in an appropriate space) are homeomorphic. That is, any arc (or life) can be continuously deformed into any other, and hence all are topologically equivalent. For example, the uppercase letters in English—C, J, L, M, N, S, U, V, W, and Z—are topologically the same. Donuts and coffee mugs as well as humans and their alimentary canals from beginning to end, to cite two other standard examples, are also homeomorphic since, were they made out of clay, one could be continuously deformed into the other, the hole in the donut and the alimentary canal becoming the handle of the mug.

Much less obvious than this topological equivalence is the Brouwer fixed point theorem, which says roughly that if you have two pieces of paper, one on top of the other, and crumple the piece on top and place it on the one on the bottom, then no matter how you crumple it, there will always be at least one point of the crumpled paper over the exact point on the bottom piece it was over before the crumpling.[2] So, in my flight-induced semi-delirium, I "reasoned" that given two contemporary people's life experiences (Jumbo's and mine), even if one person's life is very different from the other's, that is, crumpled with respect to the other's, they will have at least one common experience, which was the taxi ride I was taking.

This metaphor linking me to the taxi driver is rather strained to be sure. The topology/travel association was strengthened, however, by being twelve hours out of sync with Eastern Standard Time, the thought of Thailand's being on the other side of the earth, and another theorem in topology, the Borsuk–Ulam theorem. The latter states that every smooth function from the surface of a sphere—say, the earth—to a flat two-dimensional surface will necessarily associate a pair of points on opposite sides of the sphere to the same point on the surface. It's commonly presented in meteorological terms as saying that there will always be points on opposite sides of the earth with equal temperatures and barometric pressures.[3]

In any case Jumbo now had a car service that, even without the falloff in tourism because of the political protests at the time, barely threw off enough money for him to survive. His wife was in poor health, his children had grown, and his wife's relatives were a financial drain on him. He added matter-of-factly that when he can afford it, he still hires sex workers. Although some would abhor his nonchalance, he seemed a gentle man, maintained a sense of wistful humor about things, and as we were driving down the beach he noticed all the women and very few *farangs* (Western foreigners) and said I should have a great time since there were "so many pussies" that were unemployed. Once again, his tone was matter-of-fact and oblivious to the umbrage his remarks would cause some people.

Leaving the taxi, I noted that the flashes of similarity amid much that is foreign can be modeled, once again only quite metaphorically, using yet a different idea from topology. The technique, which I've written about in a different context, was devised by mathematician Steve Smale. Imagine a blob of spongy white bread that's been thoroughly soaked in water and pressed into a cube. Suppose further that down the middle of this cube runs a layer of grape jelly (somewhat like my grade-school sandwiches mentioned earlier). Now do the following: stretch and squeeze this cubical sandwich to twice its length, then fold it carefully back upon itself to re-form the cube. The jelly layer is now shaped like a horseshoe. Repeat this stretching, squeezing and folding many times, and you'll notice that the layer of jelly (I'm idealizing here) is soon spread throughout the dough in a most convoluted pattern. Points in the jam that were close are now distant; other points that were distant are now close. The same is true for points in the bread. Smale used this "horseshoe folding" procedure to clarify the advent of unpredictable chaos in so-called dynamical systems, of which human beings are examples.[4]

One, among the many positive effects of foreign travel, is that it does to our thinking what all the moistening, folding, and squeezing does to the bread. We hear pop songs from long ago, for example, and realize they infected some very different people on the other side of the world. Of course, we can induce this recognition in more prosaic and local ways—random shuffles on an iPod, different routes to work (if your destination is, say, eight blocks south and eight blocks west of you, combinatorics tells you there are 12,870 zigzagging ways to get there directly), or, what's so common and addictive, surfing the net in a lazy, haphazard way.

Nonetheless physical travel still beats virtual travel at revealing one's Waldo, one's self, nominal though that entity may be.

EXPERIENCING VERSUS REMEMBERING SELVES AND AUTOBIOGRAPHIES

One reason autobiographies are often vapid is that we forget the idiosyncratic details and intrigues of our own lives. The basic reason is that the minutiae of our lives are too vast and Chaitin–Kolmogorov complex for our limited brains to catalog; they're beyond our complexity horizons.

I've enjoyed teaching over the years, for example, but, unless I consciously try to dredge up specific incidents, my memories of it are largely limited to particularly lively classes, unusual students or colleagues, embarrassing moments (a gush of water from a malfunctioning sink splashing on the front of my pants just before class), warm letters or testimonials from former students, courses taught and approaches taken, and a few odd realizations. Among them: The students who finish a test first tend to be either the best or the worst. And requests for extra credit too often are poorly disguised pleas to substitute busywork for substantive coursework. Grading exams fairly, I've also learned, is an almost impossibly complex undertaking, sometimes undermined by my tendency to give As to students who come up with impressive insights about the material even if their performance on exams is less than stellar.

A similar short list of details is all I can easily recall of the physical writing of my books and columns, which seems to have dissolved into a rather amorphous memory of sitting at the computer drinking Diet Coke. As has been observed by many, a hard ass is at least as necessary as any other authorial requirement.

Most of what we hold onto of other events—an investment, adventure, relationship, long-time hobby—is the brief account of it that we often embellish and rehearse afterward and that is heavily influenced by the event's ending. The experience itself, no matter how many minutes, days, months, or years it lasted, is collapsed into a thumbnail sketch. Friends returning from vacation generally have two or three polished stories plus a brief summary that they dine out on for a few months or sometimes much longer. I've heard slight variants of their same escapades from some friends for years.

The psychologist Daniel Kahneman, whom I noted in chapter 7 and whose work on behavioral psychology earned him a Nobel prize in economics, has stressed the importance of this distinction between what he calls the experiencing self who lives through an event and the remembering self who later recalls it, and he illustrates it with an experiment involving colonoscopies that, years ago, were given to conscious patients and were quite painful.[5]

Colonoscopies that were brief but ended at a very painful point were judged afterward to be worse by patients than those that were considerably longer and fully as painful, but that ended at a point of lesser pain. That is, people who experienced the first sort of colonoscopy remembered the experience as very bad, whereas those who experienced the second sort remembered the experience as not so bad. The longer duration of pain experienced during the second sort made no difference.[6] As suggested above, the phenomenon works in the opposite direction with vacations. Short vacations that ended on a high note are often judged afterward to be better than those that were considerably longer and fully as pleasant, but that ended vexingly. More generally, we've all had interludes of one sort or another that were wonderful (or awful) but that ended on a sour (or sweet) note that disproportionately affected our remembrance of the time.

Happily, colonoscopies, if not vacations, have improved since then, but Kahneman gives many other examples and shows his point to be quite general. My interest here is in the relevance of this phenomenon to autobiographies. When they're written, whether near the height of one's powers or when one is enfeebled (embittered, uncontrollably digressive, annoyingly garrulous, crotchety, repetitive, and, yes, repetitive) will, it seems to me, have a significant impact on what one writes about oneself. It also affects whether one writes more of the heroic aspirations of one's youth, which often are simply a reckless naïveté, or of the prudent precautions of the aged self, which sometimes masquerade as mature wisdom. Pardon the simile, but the circumstances at the time of writing one's autobiography are like the aftermath of a colonoscopy.

Kahneman's way of putting it in *Thinking, Fast and Slow* is apt, albeit a little hyperbolic: "Odd as it may seem, I am my remembering self, and the experiencing self, who does my living, is like a stranger to me."[7] It is indeed odd, and I think it represents an unwarranted devaluation of the experiencing self, but that's a different issue (with Buddhist-like implications). Whatever the role and importance of our experiencing self, it is true that for psychological reasons (as well as for related ones regarding capacity—we can't remember everything with equal clarity) we tend to mentally compartmentalize our lives, and what we remember most readily, sadly even of compartments as salient as our children, are disproportionately iconic memories—their births, their first days at school, special events, ball games, snow play, parties, graduations, trips, dinners, sicknesses, achievements, and the like as well as summarizing conclusions. And we do, of course, put a lot of weight on the endings of various periods.

Even accounts of a happy, productive, and eventful life can be discolored by a painful and/or disillusioning ending. Looking back, we might see milestones as millstones. Or the other way around, of course. (I wonder how many autobiographers reread their book only to wonder at the more youthful stranger portrayed therein.) At the very least this over-weighting of endings is an added reason to hope that we check out not too much later than our cognitive and physical faculties do. Failing that, we should hope that everything possible is done to ensure a largely painfree end to life, at least for the majority of people who prefer analgesia to total lucidity. Nevertheless, since death is universal and since one knows that in a sense it's always going to end badly, all lives are personal tragedies. Happily, this feeling can be somewhat ameliorated by a consideration of one's whole life and by conceptions of self that include one's family, friends, society as a whole, even the universe, whatever it encompasses. With personal regard to the latter, the announcement of the Higgs boson on July 4, 2012, my birthday, somehow made the realization that we're all made of starstuff, mathstuff, the same stuff feel a bit more visceral.

Alas, aside from a few brief intervals, cognizance of our mortality is a monotonically increasing function.

PEAK EXPERIENCES, RECORD SETTING, AND THE PATH FROM JADE TO JADED

Quite flattering are two awards given me in recent years. In 2003 the American Association for the Advancement of Science gave me its annual award for advancing public understanding of science, and in 2013 the Joint Policy Board for Mathematics (an umbrella organization of the major mathematical societies) gave me a similar lifetime award for advancing public understanding of mathematics. I'm honored to have received these awards, but they also remind me that such recognition generally occurs late in life and that peak experiences of most other sorts (including those when one feels most intensely alive, most profoundly there) generally occur less frequently as one ages.

Let me drain the subject of aging of its personal resonance and instead focus on one aspect of it much more abstractly. To this end, imagine some random quantity. Any one will do, but let's say the number of heads in 1,000 flips of a coin. Imagine further that after flipping a coin 1,000 times and counting the number of heads, you again flip the coin 1,000 times and count the number of heads. If you do this a third time and a fourth time and keep on flipping the coin 1,000 times and counting the number of heads, how often will you achieve a record number of heads; that is, how often will you achieve a number of heads greater than the number of heads in all previous instances of 1,000 flips? Try to ignore your wrist's carpal tunnel pain.

To illustrate, let's assume that the number of heads the first ten times you flip the coin 1,000 times is 524; 496; 501; 527; 488; 499; 514; 519; 474; 531. The first result of 524 will, of course, automatically be a record number of heads, but the next record number of heads—527—occurs the fourth time you flip the coin 1,000 times, and the next record of heads after that, 531—the third record, occurs on the tenth time you flip the coin 1,000 times. That is, the sequence or records R and un-records U would in this case be RUU-RUUUUUR. Note that the records occur less and less frequently. Moreover, you probably wouldn't achieve much more than 9 Rs, 9 record numbers of heads, by the ten thousandth time you flipped the coin 1,000 times, and even by the one millionth time you flipped the coin 1,000 times you probably wouldn't achieve much more than 14 Rs, 14 record numbers of heads. Amazingly, but not coincidentally, the 9th root of 10,000 and the 14th root of 1,000,000 are approximately equal to e, which is 2.71828 . . . , the base of the natural logarithm.

Summarizing, I can say that the general relation is that if the number of times you flip a coin 1,000 times is a number N (a hundred, a million, a billion, whatever), then the number of records—that is, the number of Rs you will achieve—will be roughly equal to the natural log of N, and this a shrinking fraction of N. And the relevance of all this to biographies and the loss of excitement and verve that aging usually brings about? Interpreting record highs loosely along whatever dimension you care to mention, the mathematical fact just mentioned suggests that these high points—records, peak experiences, and the like—will occur less and less frequently as we get older. (Likewise, record floods, droughts, and fires would be occurring much less frequently were it not for global warming.)

You might say that the above argument depends on events occurring randomly, as the results of the 1,000 coin flips did, and that's true. The establishment of new records in sports, for example, occurs more frequently than that because of improvements in training and nutrition, but not that much more frequently (although the math is more complicated for such probability distributions). But most common discoveries, epiphanies, achievements (records for short) do tend to occur early rather than late—the best movie you've ever seen, the best meal you've ever had, or, of course, that mind-altering realization you had in middle school. And, though perhaps too obvious to reiterate, these records do seem to occur much more frequently to children and teenagers, much more rarely to adults who are, in this sense, almost necessarily jaded.

Of course, there are still many significant events and milestones in older people's lives, but they are not likely to be record-setting. (Ever-increasing ear size is an exception.) If people's life spans were only thirty years, this slow transition from the greenness of jade to the grayness of jaded would not be as ordinary a part of life as it is, nor would terms like "bucket list" and "midlife crisis"—fervent attempts to achieve another few records be the clichés that they are. (Picasso's quip comes to mind: "One starts to get young at the age of sixty, and then it's too late."[8]) I don't mean to mock these attempts to prolong the intensity of youth. In fact I applaud those who make them. It's just that they are somewhat predictable on mathematical grounds alone. Furthermore, it's remarkable that a notion as superficially dry (to most) as the natural logarithm of a number and the extraordinarily ubiquitous number e can have some relevance to a notion as romantic as a peak experience.

Recall that e is intimately related to the growth of your money, a strategy for choosing a spouse, the number of tries needed to amass a complete set of something, the frequency of your life's record achievements, the location of the stars in the sky, and much more. The much more includes Leonhard Euler's shiver-inducing identity, $e^{pi \cdot i} + 1 = 0$, which relates the numbers, e, pi, the imaginary number i, 1, and 0, all fundamental elements in the periodic table of mathstuff.[9]

Logarithms and exponents and the decreasing frequency of records and milestones also suggest a different sort of short biography. The idea is to include a few events and the general life situation from around the time of the subject's 1st birthday. Then in succeeding chapters do the same for the subject's 2nd birthday, 4th birthday, 8th, 16th, 32nd, and 64th birthday, focusing only on the power-of-two birthdays. So far, there doesn't seem to have been many, if any, 27th or 128th birthdays. More achievable is living to be 100 years of age, which, incidentally, is approximately pi billion seconds.

JOINING MY FATHER

I remember my father sitting and chuckling on the concrete steps outside our house in Milwaukee one humid autumn night. I asked what was funny, and he told me that he had heard Bob Buhl, a pitcher for the Milwaukee Braves during their heyday, answer a TV reporter's question about his off-season plans. "Buhl said he was going to help his father up in Saginaw, Michigan, during the winter."

My father laughed again and continued. "And when the reporter asked what Buhl's father did up in Saginaw, Buhl smiled and said, 'Nothing at all. He does nothing at all.'"[10]

My father liked this kind of story and his customary crooked grin lingered on his face. This memory was jogged recently when I was straightening out my office and found a cartoon he had sent me years ago. It showed a bum sitting happily on a park bench as a line of serious well-dressed businessmen with briefcases traipsed by him. The bum smirks and calls out, "Who's winning?" The sentiment is not all that different from that in Leo Tolstoy's short story "How Much Earth Does a Man Need?" about an ambitious, greedy man given the opportunity to possess all the land he can walk around in one day.[11] The parable suggests an answer: about 6 feet by 2 feet by 4 feet—enough for a grave. I prefer my father's cartoon.

My father was a salesman more intent on schmoozing with his customers, telling jokes, writing poetry (not all of it doggerel), and taking innumerable coffee breaks than on making a sale. He suffered a bit from logorrhea and talked to everybody who wasn't overtly hostile or uninterested. To me he was always reliable, steady, and kind—boring, old-fashioned traits perhaps, but they were the very bedrock of my early life. I remember coming out of movies as a kid and seeing him standing there, always on time to pick up my siblings and me, always with his hat askew and his warm, asymmetric smile. And I remember going to Denver every summer to visit our grandmother when we were kids and sitting in the backseat of the car late at night. We'd be driving along narrow, dark Iowa roads, and I'd see the oncoming headlights glint off the top of his bald head and feel completely safe and secure. To this day I've never felt as comfortable as a passenger with any other driver.

I can so easily picture myself listening with him to Milwaukee Braves games on hot summer nights as we sat outside on the steps of the stoop. We'd discuss the news, school, politics, the game, whether he had more hair than Warren Spahn or Lew Burdette. Sometimes we'd argue, but most of the time we talked about nothing in particular, or he told stories about his customers and their foibles. Occasionally he would take me or one of my siblings traveling with him on sales trips to such exotic locales as Wausau or Rhinelander, Wisconsin. After hours of driving on narrow roads and nighttime listening to weak AM radio stations coming from places unknown, we'd check in to a hotel and go down to the coffee shop for some chili around midnight. I remember once traveling with him to Webster City, Iowa, where our hotel room was right above the town's only traffic light and we heard cows mooing from the stopped trucks below. We ended up spending most of the night in the all-night truck stop down the block. Most of all I remember talking to him. I remember talking to him at the Rathskeller in Madison where he was sometimes mistaken for a narcotics agent because of his fedora and the ominous way he chewed on a swizzle stick; I remember talking to him in the car, at the kitchen table, over the phone, watching television. He was, as I said, a talker.

Despite appearances sometime, he was not, however, a Polly-anna. One of his favorite words was *piffle*, which he used to express resignation, disillusionment, and cynicism, as he did in his poem "The Golden Years—Bullshit." A ditty of his also comes to mind: "Remember changing their diapers, It was the beginning of their journey, But the world turns and before you know it, They have the power of attorney." Sometimes to my annoyance, he expressed skepticism about mathematics and science, not about their content but about their importance in human life. He liked a little too much a line from Samuel Beckett that I read to him once when I was in college: "All these calculations yes explanations yes the whole story false from beginning to end yes."[12] He rephrased it in a funny, earthy way that I wish I could recall.

I miss him and sometimes imagine that on my deathbed, assuming I have a death bed with a comfortable sleep number, I can say that I'm going to help my father do and be what he does and is. Nothing at all. Nothing at all.

Strike that melodramatic ending. A better, more complex, and lovely ending is. . . . Unfortunately I don't have one right now. I have only the seemingly lame mathematical knowledge that nothing at all is not the same as the empty set, which is defined to be the set that contains nothing at all. The empty set is like a nonmaterial empty bag; it contains nothing at all and has no members. Equivalently, it is the set containing all the round squares or the set containing all the whole numbers between 7 and 8. Because the difference is subtle, I'll reiterate: nothing at all and the set containing nothing at all are not the same notion. Nevertheless, I'll equivocate ever so slightly.

Although neither my father nor anyone nor anything is in the empty set, the empty set itself is procreative. A well-known fact from set theory (I'll mercifully omit the details) is that from the empty set, one can construct or generate all the whole numbers, all the real numbers, in fact, all of mathematics. If we're made out of mathstuff in one sense or another, then maybe the procreativity of the empty set can be extended. Though it contains nothing at all, the empty set might be able to generate not only all of mathematics but my father's lop-sided Cheshire grin as well.

NOTES

INTRODUCTION: WHAT'S IT ALL ABOUT?

1. Samuel Boswell, *The Life of Samuel Johnson, LL.D.* (1791); Mary Karr, *The Liars' Club: A Memoir* (Viking Penguin, 1995).
2. Stephen Jay Gould, "The Median Isn't the Message," UMassAmherst, Information Technology, accessed June 4, 2015, http://people.umass.edu/biep540w/pdf/Stephen%20Jay%20Gould.pdf.
3. Anne Tyler, *Back When We Were Grownups* (Knopf, 2001).
4. Milan Kundera, *Encounter: Essays* (HarperCollins, 2010).
5. John Allen Paulos, *Innumeracy: Mathematical Illiteracy and Its Consequences* (New York: Farrar, Straus and Giroux, 1989).

CHAPTER 1: BULLY TEACHER, CHILDHOOD MATH

1. John Allen Paulos, *Innumeracy: Mathematical Illiteracy and Its Consequences* (New York: Farrar, Straus and Giroux, 1989).
2. Sheldon Ross, *A First Course in Probability* (Macmillan, 1994).
3. John Allen Paulos, "Imagining a Hit Thriller with the Number 'e,'" ABCNews.com, accessed June 8, 2015, http://abcnews.go.com/Technology/WhosCounting/story?id=99501&page=1&singlePage=true.
4. *Wikipedia*, s.v. "Collatz conjecture," last modified June 4, 2015, accessed June 6, 2015, http://en.wikipedia.org/wiki/Collatz_conjecture.
5. Robert Kanigel, *The Man Who Knew Infinity: A Life of the Genius Ramanujan* (Simon and Schuster, 1991).
6. *Wikipedia*, s.v. "Banach-Tarski paradox," last modified May 29, 2015, accessed June 6, 2015, http://en.wikipedia.org/wiki/Banach%E2%80%93Tarski_paradox.

CHAPTER 2: BIAS, BIOGRAPHY, AND WHY
WE'RE ALL A BIT FAR-OUT AND BIZARRE

1. "Majority of Parents Abuse Children, Children Report," *Onion*, April 13, 2007, accessed June 9, 2015, http://www.theonion.com/article/majority -of-parents-abuse-children-children-report-2183.

2. Steven Wright quote, BrainyQuote, accessed June 9, 2015, http://www .brainyquote.com/quotes/quotes/s/stevenwrig102563.html.

3. Norman Stuart Sutherland, *Irrationality: The Enemy Within* (Constable, 1992).

4. Ibid.

5. Edwin A. Abbott, *Flatland: A Romance of Many Dimensions* (Cambridge, MA: Perseus, 2002).

6. *Wikipedia*, s.v. "All models are wrong," under heading "Quotations of George Box," last modified January 18, 2015, accessed June 9, 2015, http://en.wikipedia.org/ wiki/All_models_are_wrong#Quotations_of_George_Box.

7. Michael Chwe, *Jane Austen, Game Theorist* (Princeton, NJ: Princeton University Press, 2013).

8. G. H. Hardy, *A Mathematician's Apology* (Cambridge: Cambridge University Press, 1940).

9. Steven L. Goldman, Science, *Technology, and Social Progress*, Google Books, accessed June 23, 2015, https://books.google.com/books?id=MuPbmJ_pzmcC&pg =PA150&lpg=PA150&dq=from+such+cloistered+clowning+the+world+sickens&source =bl&ots=LsfEuqF0Ah&sig=HpQSdvzpJWDtvnfDJ9Tehb9bums&hl=en&sa=X&ei =RFWJVfCTKNLpoATyub6oDw&ved=0CB8Q6AEwAA#v=onepage&q=from%20such %20cloistered%20clowning%20the%20world%20sickens&f=false.

CHAPTER 3: AMBITION VERSUS NIHILISM

1. Paul Halmos, *Naïve Set Theory* (Van Nostrand, 1960).

2. Ibid.

3. Ibid.

4. Laurence Sterne, *The Life and Opinions of Tristram Shandy* (Wordsworth Editions, 1996).

5. Tetyana Butler, "Paradox of Tristram Shandy," Suitcase of Dreams, accessed June 9, 2015, http://www.suitcaseofdreams.net/Tristram_Shandy.htm.

6. John Allen Paulos, *A Mathematician Reads the Newspaper* (New York: Basic Books, 2013).

7. Thomas Nagel, "The Absurd," *Journal of Philosophy*, University of Kentucky

online, accessed June 9, 2015, https://philosophy.as.uky.edu/sites/default/files/ The%20Absurd%20-%20Thomas%20Nagel.pdf.

8. Groucho Marx quote, Genius Quotes, accessed June 23, 2015, http://genius .com/2190673.

9. John Allen Paulos, *Irreligion: Why the Arguments for God Just Don't Add Up* (New York: Farrar, Straus and Giroux, 2008).

10. Douglas Hofstadter, *I Am a Strange Loop* (New York: Basic Books, 2007).

11. Ibid.

12. Ibid.

13. Ibid., pp. 291–93.

14. Herman Melville quote from *Moby Dick*, *Bibliographing*, accessed June 9, 2015, http://www.bibliographing.com/2010/07/14/i-promise-nothing-complete-because -any-human-thing-supposed-to-be-complete-must-for-that-very-reason-infallibly-be -faulty/.

15. Robert Nozick, *Anarchy, State, and Utopia* (New York: Basic Books, 1974).

CHAPTER 4: LIFE'S SHIFTING SHAPES

1. George Lakoff and Rafael Nuñez, *Where Mathematics Comes From: How the Embodied Mind Brings Mathematics into Being* (New York: Basic Books, 2001).

2. Carl Japikse and Franklin Japikse, *Fart Proudly: Writings of Benjamin Franklin You Never Read in School* (Marble Hill, GA: Enthea Press, 1990).

3. Tolstoy quote from *War and Peace*, Mathematical Fiction, accessed June 10, 2015, http://kasmana.people.cofc.edu/MATHFICT/mfview.php?callnumber=mf117.

4. Ibid.

5. Thich Nhat Hanh, "Clouds in Each Paper," *Awakin.org*, March 25, 2002, accessed June 10, 2015, http://www.awakin.org/read/view.php?tid=222.

6. Peter Dizikes, "When the Butterfly Effect Took Flight, *MIT Technology Review*, February 22, 2011, accessed June 10, 2015, http://www.technologyreview.com/ article/422809/when-the-butterfly-effect-took-flight/.

7. John Allen Paulos, "We're Measuring Bacteria with a Yardstick," Opinion, *New York Times*, November 22, 2000, accessed June 10, 2015, http://www.nytimes .com/2000/11/22/opinion/22PAUL.html?ex–1078549200&en=1bbd1e85c6ec0af5 &ei=5070.

8. John Allen Paulos, "How Florida's Chief Judge Misquoted Me," *Philadelphia Daily News*, December 11, 2000, accessed June 10, 2015, http://articles.philly .com/2000-12-11/news/25578951_1_al-gore-popular-vote-voter-news-service.

9. Ibid.

10. Jeff Greenfield, *Oh, Waiter! One Order of Crow!: Inside the Strangest Presidential Election Finish in American History* (Putnam Adult, 2001).

11. Jack Copeland, *Turing: The Pioneer of the Information Age* (Oxford: Oxford University Press, 2014).

CHAPTER 5: MOVING TOWARD THE UNEXPECTED MIDDLE

1. Heraclitus quote, YourDictionary, accessed June 11, 2015, http://quotes .yourdictionary.com/author/heraclitus/.

2. George Carlin quote, BrainyQuote, accessed June 11, 2015, http://www .brainyquote.com/quotes/quotes/g/georgecarl391403.html.

3. Franz Kafka quote from *The Zürau Aphorisms*, Brilliant.Life.Quotes., accessed June 11, 2015, http://www.brilliantlifequotes.com/inspirational/franz -kafka-quotes-life-suffering-religion/.

4. John Allen Paulos, *Once Upon a Number: The Hidden Mathematical Logic of Stories* (New York: Basic Books, 1998).

5. John Allen Paulos, *Mathematics and Humor: A Study of the Logics of Humor* (Chicago: University of Chicago Press, 1980).

6. Anne Tyler, *Back When We Were Grownups* (Ballantine Books, 2002).

7. *Wikipedia*, s.v. "Reminiscence bump," last modified April 9, 2015, accessed June 11, 2015, http://en.wikipedia.org/wiki/Reminiscence_bump.

8. Daniel Kahneman, *Thinking, Fast and Slow* (New York: Farrar, Straus and Giroux, 2011).

9. Benford's law, Math Pages, accessed June 11, 2015, http://www.mathpages .com/home/kmath302/kmath302.htm.

10. Daniel Schacter, *The Seven Sins of Memory* (Boston: Houghton Mifflin, 2001).

CHAPTER 6: PIVOTS—PAST TO PRESENT

1. Tzetzka Ilieva, "Sofia Kovalevskaya, Mathematician," *History's Women*, accessed June 23, 2015, http://www.historyswomen.com/moregreatwomen/Sofia Kovalevskaya.htm.

2. Alice Munro, *Too Much Happiness* (New York: Vintage, 2010).

3. Sofia Kovalevskaya, *Nihilist Girl* (New York: Modern Language Association of America, 2002).

4. Greg Chaitin, *Algorithmic Information Theory (Cambridge Tracts in Theoretical Computer Science)* (Cambridge: Cambridge University Press, 1988).

5. Claude Shannon, *The Mathematical Theory of Communication* (Urbana: University of Illinois Press, 1949).

6. George Johnson, *Fire in the Mind* (New York: Knopf, 1995).

7. Bertrand Russell, letter to the author, in *Autobiography of Bertrand Russell* (London: George Allen and Unwin version, 1975).

○

CHAPTER 7: ROMANCE AMONG THE TRANS-HUMANS
AND US CIS-HUMANS

1. The 37 percent figure and the optimal way to find your best partner: *Wikipedia*, s.v. "Secretary problem," last modified January 13, 2015, accessed June 14, 2015, http://en.wikipedia.org/wiki/Secretary_problem.

2. Alain de Botton, "On the Madness and Charm of Crushes," *Philosopher's Mail*, accessed June 14, 2015, http://thephilosophersmail.com/relationships/on-the-madness-and-charm-of-crushes/.

3. John Allen Paulos, "The Advanced Metrics of Attraction," *New York Times*, July 14, 2014, accessed June 14, 2015, http://www.nytimes.com/2014/07/15/science/the-advanced-metrics-of-attraction.html?_r=0

4. Daniel Kahneman, *Thinking, Fast and Slow* (New York: Farrar, Straus and Giroux, 2013).

5. Ibid.

6. *Wikipedia*, s.v. "Zeno's paradoxes," last modified June 15, 2015, accessed June 23, 2015, https://en.wikipedia.org/wiki/Zeno's_paradoxes.

CHAPTER 8: CHANCES ARE THAT CHANCES ARE

1. Ecclesiastes "Time and chance happeneth to all."

2. Sheldon Ross, *A First Course in Probability* (Macmillan, 1994).

3. Arthur C. Clarke, *Rendezvous with Rama* (New York: Bantam Books, 1973).

4. John Allen Paulos, *Innumeracy: Mathematical Illiteracy and Its Consequences* (New York: Farrar, Straus and Giroux, 1989).

5. Douglas Hofstadter, *Gödel, Escher, and Bach: An Eternal Golden Braid* (New York: Basic Books, 1999).

6. John Allen Paulos, *A Mathematician Reads the Newspaper* (New York: Basic Books, 2013).

7. John Allen Paulos, *I Think, Therefore I Laugh* (New York: Columbia University Press, 1985).

8. John Allen Paulos, archive of "Who's Counting" columns, ABCNews, accessed June 15, 2015, http://abcnews.go.com/search?searchtext=%22John%20allen%20paulos%22#.

9. John Allen Paulos, *Irreligion: A Mathematician Explains Why the Arguments for God Just Don't Add Up* (Hill and Wang, 2009).

CHAPTER 9: LIVES IN THE ERA OF NUMBERS AND NETWORKS

1. Stephen Wolfram, *A New Kind of Science* (Wolfram Media, 2002).

2. Seth Stephens-Davidowitz, "Searching for Sex," Sunday Review, *New York Times*, January 24, 2015, accessed June 20, 2015, http://www.nytimes.com/2015/01/25/opinion/sunday/seth-stephens-davidowitz-searching-for-sex.html.

3. Wolfram, *A New Kind of Science*.

4. Ibid.

5. Bruce Goldman, "Mathematics or Memory: Study Charts Collision Course in Brain," Stanford Medicine News Center, accessed June 16, 2015, http://med.stanford.edu/news/all-news/2012/09/mathematics-or-memory-study-charts-collision-course-in-brain.html.

6. Marcel Proust, *In Search of Lost Time* (Modern Library, 1981).

7. Pablo Picasso, "Statement to Marius de Zayas," 1923, Columbia University Media Center for Art History, accessed June 20, 2015, http://www.learn.columbia.edu/monographs/picmon/pdf/art_hum_reading_49.pdf.

8. *Wikiquote*, s.v. "Blaise Pascal," last modified May 31, 2015, accessed June 16, 2015, http://en.wikiquote.org/wiki/Blaise_Pascal.

9. John Allen Paulos, Twitter account, http://www.twitter.com/johnallenpaulos.

10. For more on Erdős numbers, see *Wikipedia*, s.v. "Erdős number," last modified May 30, 2015, accessed June 16, 2015, http://en.wikipedia.org/wiki/Erd%C5%91s_number.

11. David Foster Wallace, *Everything and More: A Compact History of Infinity* (Norton, 2003).

12. *Wikipedia*, s.v. "Friendship paradox," last modified March 13, 2015, accessed June 16, 2015, http://en.wikipedia.org/wiki/Friendship_paradox; see also John Allen Paulos, "Why You're Probably Less Popular Than Your Friends, *Scientific American*, January 18, 2007, accessed June 16, 2015, http://www.scientificamerican.com/article/why-youre-probably-less-popular/.

13. Alexander Herzen, *My Past and Thoughts* (Oakland: University of California Press, 1982).

14. Louis Menand, "Everybody's an Expert," *New Yorker*, December 5, 2005, accessed June 20, 2015, http://www.newyorker.com/magazine/2005/12/05/everybodys-an-expert.

15. *Wikipedia*, s.v. "The Hedgehog and the Fox," last modified February 3, 2015, accessed June 20, 2015, https://en.wikipedia.org/wiki/The_Hedgehog_and_the_Fox.

CHAPTER 10: MY STOCK LOSS, HYPOCRISY, AND A CARD TRICK

1. *Wikipedia*, s.v. "Bernie Ebbers," last modified May 14, 2015, accessed June 17, 2015, https://en.wikipedia.org/wiki/Bernard_Ebbers.

2. John Allen Paulos, "My Lowest Ebb(ers)," College of Science and Technology, Temple University, March 1, 2005, https://math.temple.edu/~paulos/ebbers.html.

3. *Wikipedia*, s.v. "Bernie Ebbers."

4. John Allen Paulos, *A Mathematician Plays the Stock Market* (New York: Basic Books, 2003).

5. John Allen Paulos, *Innumeracy: Mathematical Illiteracy and Its Consequences* (New York: Farrar, Straus and Giroux, 1989).

6. *Wikipedia*, s.v. "Gettier problem," last modified June 15, 2015, accessed June 17, 2015, http://en.wikipedia.org/wiki/Gettier_problem.

7. Definition of *hypocrisy*, Free Dictionary, accessed June 17, 2015, http://www.thefreedictionary.com/hypocrisy.

8. *Wikipedia*, s.v. "Church–Turing thesis," last modified June 4, 2015, accessed June 17, 2015, https://en.wikipedia.org/wiki/Church%E2%80%93Turing_thesis.

9. *Wikipedia*, s.v. "Gottfried Wilhelm Leibniz," last modified June 14, 2015, accessed June 17, 2015, https://en.wikipedia.org/wiki/Gottfried_Wilhelm_Leibniz.

10. *Wikipedia*, s.v. "Boolean satisfiability problem," last modified May 15, 2015, accessed June 17, 2015, http://en.wikipedia.org/wiki/Boolean_satisfiability_problem.

11. *Wikipedia*, s.v. "Felix Klein," last modified May 25, 2015, accessed June 17, 2015, https://en.wikipedia.org/wiki/Felix_Klein.

12. "The Kruskal Count," Mathematical Card Tricks, American Mathematical Society, accessed June 17, 2015, http://www.ams.org/samplings/feature-column/fcarc-mulcahy6.

13. John Allen Paulos, *Once Upon a Number: The Hidden Mathematical Logic of Stories* (New York: Basic Books, 1998).

14. *Wikipedia*, s.v., "Markov chain," last modified June 16, 2015, accessed June 17, 2015, https://en.wikipedia.org/wiki/Markov_chain.

CHAPTER 11: BIOGRAPHIES: VERSTEHEN OR SUPERFICIAL

1. Garrison Keillor, "Mark Twain's Riverboat Ramblings," Sunday Book Review, *New York Times*, December 16, 2010, accessed June 20, 2015, http://www.nytimes.com/2010/12/19/books/review/Keillor-t.html?pagewanted=all.

2. Paul Auster quote, goodreads, accessed June 20, 2015, https://www.goodreads.com/quotes/41926-every-life-is-inexplicable-i-kept-telling-myself-no-matter.

3. Rebecca West quote, Quotes.net, accessed June 20, 2015, http://www.quotes .net/quote/18882.

4. *Wikipedia*, s.v. *Verstehen*, last modified October 18, 2014, accessed June 20, 2015, http://en.wikipedia.org/wiki/Verstehen.

5. Donald E. Knuth, "The Complexity of Songs," fivedots.coe.psu.ac.th, accessed June 20, 2015, http://fivedots.coe.psu.ac.th/Software.coe/242-535_ADA/Background/ Readings/ knuth_song_complexity.pdf.

6. KC and the Sunshine Band, "That's the Way (I Like It)," MetroLyrics, accessed June 20, 2015, http://www.metrolyrics.com/thats-the-way-i-like-it-lyrics-kc-and -the-sunshine-band.html.

7. "Berry Paradox," WolframMathWorld, accessed June 20, 2015, http://math world.wolfram.com/BerryParadox.html.

8. "The Repugnant Conclusion," Stanford Encyclopedia of Philosophy, September 9, 2010, accessed June 20, 2015, http://plato.stanford.edu/entries/ repugnant-conclusion/.

9. Thomas Nagel, "What Is It Like to be a Bat?" *Philosophical Review* 83, no. 4 (October 1974): 435–50, accessed June 20, 2015, http://organizations.utep.edu/ portals/1475/nagel_bat.pdf.

10. Ibid.

11. Ibid.

12. Ibid.

13. *Wikiquote*, s.v. "Robert Frost," last modified February 6, 2015, accessed June 20, 2015, https://en.wikiquote.org/wiki/Robert_Frost.

14. Brian Skinner on Gompertz, "Your Body Wasn't Built to Last: A Lesson from Human Mortality Rates, *Gravity and Levity*, July 8, 2009, accessed June 20, 2015, https://gravityandlevity.wordpress.com/2009/07/08/your-body-wasnt-built -to-last-a-lesson-from-human-mortality-rates/.

15. Robert Krulwich, "Am I Going to Die This Year? A Mathematical Puzzle," Krulwich Wonders, NPR, January 8, 2014, accessed June 20, 2015, http://www .npr.org/sections/krulwich/2014/01/08/260463710/am-i-going-to-die-this-year -a-mathematical-puzzle.

16. David Quammen, "Lives of the Cells," Sunday Book Review, *New York Times*, September 6, 2013, accessed June 20, 2015, http://www.nytimes.com/2013/09/08/ books/review/george-johnsons-cancer-chronicles.html.

17. Ibid.

CHAPTER 12: TRIPS, MEMORIES, AND BECOMING JADED

1. John Allen Paulos, *Beyond Numeracy* (Knopf, 1991).

2. William Chinn and N. E. Steenrod, *First Concepts of Topology: The Geom-*

etry of Mappings of Segments, Curves, Circles, and Disks (Mathematical Association of America, 1966).

3. Francis Su et al., "Borsuk–Ulam Theorem," *Math Fun Facts*, accessed June 20, 2015, https://www.math.hmc.edu/funfacts/ffiles/20003.7.shtml.

4. *Wikipedia*, s.v. "Horseshoe map," last modified May 28, 2014, accessed June 20, 2015, http://en.wikipedia.org/wiki/Horseshoe_map.

5. Daniel Kahneman, *Thinking, Fast and Slow* (New York: Farrar, Straus and Giroux, 2013).

6. Ibid.

7. Ibid.

8. Picasso quote, Milestone birthday quotations, write-out-loud.com, accessed June 20, 2015, http://www.write-out-loud.com/milestone-birthday-quotations.html.

9. *Wikipedia*, s.v. "Euler's identity," last modified June 2, 2015, accessed June 20, 2015, http://en.wikipedia.org/wiki/Euler%27s_identity.

10. *Wikipedia*, s.v. "Bob Buhl," last modified February 1, 2015, accessed June 20, 2015, http://en.wikipedia.org/wiki/Bob_Buhl.

11. Leo Tolstoy, "How Much Earth Does a Man Need?" Literature Network, accessed June 20, 2015, http://www.online-literature.com/tolstoy/2738/.

12. Beckett quote, Notable Quotes, accessed June 20, 2015, http://www.notable-quotes.com/b/beckett_samuel.html.

INDEX

algorithms, 92, 103, 104, 152
  and Alan Turing, 152
American Association for the Advancement of Science, 182–83
Anarchy, State, and Utopia (Nozick), 60–61
anti-memoir, 12
Asimov, Isaac, 102–103
Austen, Jane
  and game theory, 47
autobiographers, 36–37, 72, 182
autobiographies, 38, 55, 57, 66, 84, 144, 179
  and Benford's law, 91–92
  differences between biographies, 79
  and memory, 11, 90–91, 181

Back When We Were Grownups (Tyler), 88
Bacon, Kevin. *See under* six degrees of separation
Banach space, 46–47
Banach–Tarski theorem, 33
Bayes' theorem, 40, 107, 111–12. *See also* statistics, Bayesian
Beckett, Samuel, 187
Being and Nothingness (Sartre), 99
Bellow, Saul, 132
bell-shaped normal distribution (Gauss), 45–46
Benford's law, 89–92
  and autobiographies, 91–92
  and memories, 89–90
Berry paradox, 167

and Bertrand Russell, 167
  See also paradox
Beyond Numeracy (Paulos), 176–77
bias, 37, 65, 88
  and biographies, 11–12, 75, 77
  confirmation, 38, 146
  sampling, 111
  selection, 29
biographers, 36–37, 40, 72, 75, 101, 102, 142, 149
  neutrality of, 37–38
biographies, 18, 37–38, 55–56, 57, 61, 88–89, 104, 106, 112, 127, 141–42, 148, 150, 153, 162, 168
  attitude toward, 11–12, 38, 41, 91–92
  and bias, 11–12, 75, 77
  and coincidence, 121
  description of, 13, 18, 35, 71, 98
  differences between autobiographies, 79
  and Facebook (Wolfram), 130, 132
  and history, 148–49
  inadequacies of, 16–17, 36–37, 40, 55, 77–78, 147, 161
  and life's progression, 64–65, 99
  and Linda problem, 76
  and mathematics, 14–16, 46, 48, 65–66, 147–48, 149, 158–59, 184 85
  and notion of "I," 143–44
  and paradox, 60
  and roboromance, 105–108

and romance, 110–12
of Sofia Kovalevsky, 93–95
and statistics, 37–40, 68
and Texas sharpshooter fallacy, 74–76
and Twitter, 133–35, 139
and Verstehen, 162–65, 168
Black–Scholes formula, 47
Boolean satisfiability problem, 153
Borsuk–Ulam theorem, 178
Boswell, James (*The Life of Samuel Johnson, LL.D.*), 12
Botton, Alain de ("On the Madness and Charm of Crushes"), 110
Box, George, 46
Brouwer fixed point theorem, 177
Buddhism, 55
Bush, George W.
butterfly effect in election of, 72–73
butterfly effect, 70, 71, 72–73, 143

calculus, 67–68, 71
in *War and Peace* (Tolstoy), 68
cancer
and "cops and criminals" theory of the immune system, 173–74
inevitability of (Johnson), 173–74
Cantor, Georg, 49. *See also* real numbers
cardinality, 49–50
cardinal numbers, 51. *See also* numbers, types of
card tricks
mathematical (Kruskal), 155–57
and probability, 117
and probability theory, 28
careenium, 58
central limit theorem, 45–46

Chaitin, Greg (*The Limits of Mathematics*), 104, 166
Chaitin–Kolmogorov complex, 179
chaos, 94, 179
chaos theory, 69, 94
chimera, 55
Church, Alonzo, 152
Chwe, Michael (*Jane Austen, Game Theorist*), 47
cis-human, 105–107
Clarke, Arthur C., 121
coincidences, 81–82, 119, 120, 121, 126
and probability, 116–17, 119–21, 126
Collatz conjecture, 31
combinatorics, 42, 179
confirmation bias, 38, 146
conjunction fallacy, 76
continuum hypothesis, 50–51
Copeland, Jack, 75
"cops and criminals" theory of the immune system, 173–74
cryptography, 75

de Botton, Alain. *See* Botton, Alain de
decimal numbers, 30, 49, 52. *See also* numbers, types of
Descartes, René, 125
Devlin, Keith, 26–27
dimensions, measurable, 41–44
divine intervention, 117–18
Duhem, Pierre, 97
Dyson, Freeman, 169

Ebbers, Bernie, 146
economics
changes of, 94
ideologies of, 153
modern, 70

economic systems, 69, 71
empty set, 188
entropy, 174
Entscheidungsproblem, 152
equation
    algebraic, 30
    differential, 48, 93–94
    Laplace's heat and wave equations, 93
    Navier–Stokes equations for fluid dynamics, 93
    nonlinear, 69, 143
    summarizing, 65
    "Theory of Partial Differential Equations, The" (Kovalevsky), 94
Erdős, Paul, 136
Erdős number, 136
essential self, 55
Euler, Leonhard, 138, 185
    Euler's identity, 185
expected value of a quantity, 27–28

Facebook, 94, 130, 132, 139
    and biography (Wolfram), 130, 132
    See also social media; Twitter
"Fart Proudly" (Franklin), 69, 72
Feld, Scott
    and networks, 138
Feynman, Richard
    and perspective on historical figures, 17
Fire in the Mind (Johnson), 103
fluid dynamics. *See* Navier–Stokes equations for fluid dynamics
Fourier series, 52, 65
fractals, 164, 167
fractions, 49, 51, 134, 184. *See also* numbers, types of

Franklin, Benjamin ("Fart Proudly"), 65, 72
Fundamental Confusion of Coincidences, 81–82, 119, 120, 121, 126. *See also* probability

Galileo
    and infinite numbers, 51
game theory, 108
    and Jane Austen, 47
Gardner, Martin, 29–30, 86
Gauss, Carl Friedrich
    bell-shaped normal distribution, 45–46
Gettier, Edmund, 149
Gettier's paradox, 150
Gödel, Escher, and Bach (Hofstadter), 123
Gödel, Kurt, 33, 59, 167
    and Gödelian constraints, 60
    incompleteness theorem, 60, 167
    and *Moby Dick* (Melville), 60
Gompertz, Benjamin, 172–73
Gompertz's law, 173
    and "cops and criminals" theory of the immune system, 174
    and mortality, 173
Gone with the Wind
    analysis of love story in, 48, 93
    See also Volterra's prey-predator systems
googol, 134
Gould, Stephen Jay
    "The Median Isn't the Message," 12

Hardy, G. H. (*A Mathematician's Apology*), 31, 48
heat and wave equations. *See* Laplace's heat and wave equations

hedonism, 61, 82
  and Henry Miller, 55
Heraclitus, 80, 135
Higgs boson, 169, 182
Hilbert, David
  infinite hotel, 51
History of Western Philosophy, The
  (Russell), 100
Hofstadter, Douglas (*Gödel, Escher, and
  Bach, I Am a Strange Loop*), 58–59, 123
    theory of simms and simmballs,
    58–59, 60–61, 173–74
Hume, David, 40, 55
humor, 3, 60, 86, 113
  connection with mathematics, 14,
    86, 103
hypercube, 43–44
hypocrisy, 151–53
  definition of, 152
  and sexual morality, 153

I Am a Strange Loop (Hofstadter), 58
"I" symbols. *See under* symbol
I Think, Therefore I Laugh (Paulos), 124
identity
  changes in, 90
  and gender, 42
  and groups, 137
  mistaken, 148–50
  personal, 57, 88, 133
  of self, 56
  social, 87–88
immortality, 51, 53
incidence matrix, 137
infinite hotel (Hilbert), 51
infinite numbers
  and Galileo, 51
  properties of, 52–53
  See also numbers, types of

infinite sets, 49–50, 51
infinity, 49–51, 53
innumeracy, 20, 122, 127
  test for, 21
Innumeracy (Paulos), 15, 20, 24, 100–
  101, 122, 125, 126
integers, 49–50. *See also* numbers,
  types of
Irreligion (Paulos), 56, 126

Johnson, George (*Fire in the Mind*),
  103, 174
  "cops and criminals" theory of the
    immune system, 173–74
  and inevitability of cancer, 174

Kahneman, Daniel (*Thinking, Fast and
  Slow*), 90, 111, 180–81
  and system 2 thinking, 112
Karr, Mary (*Liars' Club*), 12
Klein, Felix
  and geometrical figures, 154
Knuth, Donald
  and complexity of songs, 166
Kolmogorov, Andrey
  Chaitin–Kolmogorov complex, 179
  and informal notion of complexity,
    166
Kovalevsky (Kovalevskaya), Sofia,
  93–95
  and biography, 93–95
  differential equation, 94
  "The Theory of Partial Differential
    Equations," 94
Kruskal, Martin
  mathematical card trick, 154–58,
    172
Kundera, Milan
  and definition of biography, 18

Lake Wobegon effect
    and biography, 38
Lakoff, George (with Nuñez, *Where Mathematics Comes From*), 63
Laplace's heat and wave equations, 93
Liars' Club (Karr), 12
Life and Opinions of Tristram Shandy, Gentleman, The (Sterne)
    and paradox of Tristram Shandy (Russell), 52–53
Life of Samuel Johnson, LL.D., The (Boswell), 12
life span, 184
    average, 57–58, 65–66
    and Gompertz's law, 172–74
    statistical distribution of, 12–13
Limits of Mathematics, The (Chaitin), 166
Linda problem, 76
logarithms, 28, 30, 52, 109, 184, 185
Lorenz, Edward
    and butterfly effect, 70
Löwenheim–Skolem theorem, 89

Markov chains, 26, 159
mathematical certainty, 23–24
mathematical pedagogy, 23–24, 25, 123
Mathematica software (Wolfram), 129
Mathematician Plays the Stock Market, A (Paulos), 15, 147
Mathematician Reads the Newspaper, A (Paulos), 15, 53–54, 124–25
Mathematician's Apology, A (Hardy), 48
mathematics
    and algorithms, 103
    and baseball, 24–25
    basic, 122
    and bell-shaped distribution, 36
    and biography, 14–16, 46, 48, 65–66, 147–48, 149, 158–59, 184–85
    connection with humor, 14, 86, 103
    in daily life, 46–47
    and multiplication principle, 42
    Platonic concept of, 71–72
    and probability theory, 28–29, 71, 159
    proto-mathematics, 64
    and statistics, 46
    and Tolstoy, 67–68
    and use of calculators, 126
Mathematics and Humor (Paulos), 86, 103
mathism, 46, 48
mathstuff, 14, 67, 71, 105, 115, 155, 174, 182, 185, 188
Maxwell's demon, 71, 103
Maxwell's electromagnetic theory, 93
"Median Isn't the Message, The" (Gould), 12
Melville, Herman (*Moby Dick*)
    and Gödelian constraints, 60
memories
    accuracy of, 11–12
    and Benford's law, 89–92
    and biographies, 89–90
    childhood, 79–80
    primal, 79
    reminiscence bump, 90, 155
memory
    and autobiographies, 11, 90–91, 181
    and numbers, 92
meta-levels, 53–54
meta-memoir, 12
Miller, Henry
    and hedonism, 55

mind patterns, 60

miracles, 40, 117

Moby Dick (Melville)
  and Gödelian constraints, 60

mortality, 137, 172, 182
  annual rate of, 173
  and Gompertz's law, 172–73

multiplication principle, 42

Munchausen syndrome, 77

Munro, Alice (*Too Much Happiness*), 94

Nagel, Thomas, 54, 168, 169
  "What Is It Like to Be a Bat?" 168

narcissism, 57, 72

natural selection, 121

Navier–Stokes equations for fluid
  dynamics, 93

neuronal self, 58. *See also* symbol; "I"
  symbols (neuronal self)

New Kind of Science, A (Wolfram),
  129–30

Newton's laws of motion, 93

Nhat Hanh, Thich, 69–70

nihilism, 53, 54

nonlinear dynamics, 69–70, 71, 72

nonlinear interactions, 70

nonlinear scaling, 140, 142

Nozick, Robert (*Anarchy, State, and
  Utopia*), 60–61

numbers, types of, 49–51
  author's love of, 19
  defining people in terms of,
    42–43, 44, 71
  and memory, 92
  statements about, 59
  See also cardinal numbers;
    decimal numbers; fractions;
    infinite numbers; integers;
    random numbers; rational

numbers; real numbers; whole
  numbers

numerical estimation, 20–22

Nuñez, Rafael (with Lakoff, *Where
  Mathematics Comes From*), 63

obituaries, 65
  as biographies, 66

Once Upon a Number (Paulos), 17, 85, 157

"On the Madness and Charm of
  Crushes" (Botton), 110

order-of-magnitude estimates, 19–20

paradox
  Berry paradox, 167
  and biography, 60
  follower paradox, 139
  Gettier's paradox, 150
  of Tristram Shandy, 52–53
  Zeno's paradox, 51–52, 114

parallel sequences of events, 118

Pareto's 80–20 principle, 92

Parfit, Derek, 167–68

Parvizi, Josef, 131

Pascal, Blaise, 134

patterns, 14, 59, 60, 71, 74, 119, 121

Peace Corps
  author's experience in, 18, 34, 99,
    109

personal identity, theory of, 57–58,
  88, 133

physical entropy, 103
  and Wojciech Zurek, 103–104

Poincaré, Henri, 53, 94

Poincaré conjecture, 94

Poisson statistical distribution, 40. *See
  also* statistical distributions

polls, accuracy of, 35–36

preferences, 42, 44, 45

probabilistic analysis, 117
probabilistic misunderstandings, 38
probability, 120–21, 146–47
    and card tricks, 117
    and coincidence, 116–17, 119–21,
        126
    and coin flips, 119–20
    and rolling dice, 25–29
    and statistics, 64
    See also Fundamental Confusion
        of Coincidences
probability theory, 28–29, 71, 159
    and Bayes' theorem, 40
    and card tricks, 28
    and Markov chains, 159

quantitative literacy, 20, 125

Ramanujan, Srinivasa, 31, 93
randomness, 50, 119, 121, 135, 179, 183
    geometric random variables, 28
    occurrence of events, 28, 146,
        158, 184
    selection, 30, 70, 91, 111
    sequences, 119, 156, 166
    and simms and simmballs, 58
random numbers, 30, 157. *See also*
    numbers, types of
rational numbers, 49, 51. *See also*
    numbers, types of
real numbers, 49–51, 188. *See also*
    numbers, types of
Riemann, Bernhard, 48
Russell, Bertrand (*The History of
    Western Philosophy*), 52–53, 100
    author's experience with, 100–101
    and Berry paradox, 167
    and paradox of Tristram Shandy,
        52–53

Sagan, Carl
    comparison of mathstuff and star-
        stuff, 14
sampling, 38, 112
    bias, 111
Sartre, Jean-Paul (*Being and Nothing-
    ness*), 99
scale
    effect of, 141
    and predictability, 140–44
Schacter, Daniel, 92
scientism, 46, 48
set theory, 49–50, 188
Shandy, Tristram. *See* paradox, of
    Tristram Shandy; *see under* Russell,
    Bertrand
Shannon, Claude, 104
siblings, development of, 72
simms and simmballs (Hofstadter),
    58–59, 60–61, 173–74
    and randomness, 58, 173
six degrees of separation, 135–36
    and Erdős number, 136
    and Kevin Bacon, 136
Skinner, Brian, 173
Smale, Steve, 178–79
Snow, C. P., 104–105
social media, 123, 130, 133, 139. *See
    also* Facebook; Twitter
starstuff. *See under* Sagan, Carl
statistical distributions, 45
    of life spans, 12
    See also Poisson statistical
        distribution
statistical failings, 38
statistical overfitting, 68
statistics, 46
    baseball, 24–25
    Bayesian, 112

and biography 36–40, 68
and de Botton's thesis, 110
and distributions, 12–13
and probability, 64
Stern, Laurence (*The Life and Opinions of Tristram Shandy, Gentleman*), 52–53
stock market, 121
    and biographies, 149
    and predictability, 145–47
symbol, 58–59, 60
    "I" symbols (neuronal self), 59
    for simmballs, 58
symbolic patterns, 59
symbolic thought, 58, 59

teleportation
    in thought experiments, 57–58
Tetlock, Philip
    and prediction, 144
Texas sharpshooter fallacy, 74–76
thermodynamics, 103–104, 174
Thinking, Fast and Slow (Kahneman), 111, 181
Tolstoy, Leo (*War and Peace*), 68–69, 132
    and mathematics, 67–68
Too Much Happiness (Munro), 94
topology, 176–77
    association with travel, 177–78
    and Borsuk–Ulam theorem, 178
trans-human, 105–107
Turing, Alan, 75, 105
    and algorithms, 152
Turing machines, 75, 105
Twain, Mark
    and Halley's comet, 119
    quotes, 134, 161–62
Twitter, 133–39. *See also* Facebook; social media

Tyler, Anne (*Back When We Were Grownups*), 15, 88
Type I error, 39
Type II error, 39

ultrafilters, 50

vector analysis, 47
Verstehen, 162–66, 168
    and biographies, 162–65, 168
Volterra's prey-predator systems, 93. See also *Gone with the Wind*

War and Peace (Tolstoy)
    calculus in, 68
Watts–Strogatz model, 135
Weierstrass, Karl, 94
Wells, Charles T., 73
"What Is It Like to Be a Bat?" (Nagel), 168
Where Mathematics Comes From (Lakoff and Nuñez), 63
whole numbers, 49, 51, 89, 188
    and Berry paradox, 167
    and Collatz conjecture, 31
    See also numbers, types of
Wittgenstein, Ludwig, 13
Wolfram, Stephen (*A New Kind of Science*), 129–30, 132
    and Facebook, 130
    Mathematica software, 129
    and the social universe, 130
    Wolfram Alpha, 129–30
WorldCom, 145–47

Zeno's paradox, 51–52, 114
Zurek, Wojciech
    and physical entropy, 103–104